AF732485

IS CITES PROTECTING WILDLIFE?

This book assesses the Convention on International Trade in Endangered Species of Wild Fauna and Flora (CITES), examining both implementation and compliance.

Humans are causing a biodiversity crisis, where 1 million species are facing extinction. Species are dying, in no small part, because they are overexploited, poached, and trafficked, and CITES is the main international instrument designed to protect traded wildlife. Does the state of the world's species mean CITES is failing? This book explores the implementation of and compliance with CITES by all 183 member countries. It is imperative we know the nature and extent of the implementation of and compliance with CITES legislation in all parties to fully understand the impact of legal and illegal trade on species survival. Through extensive legislative content analysis, a Delphi iterative survey, and semi-structured interviews, this is the first book to share empirical research about CITES implementation and compliance. This book contains a comprehensive analysis of the state of CITES, what is done well, what could be done better, and what the future might bring to try to curtail the slide of the world's wildlife into extinction. By identifying lessons learned in relation to CITES legislation, implementation, and compliance, this book provides hard evidence to member countries as to how their own practice can be improved.

This timely book will be essential reading for students and academics interested in wildlife law, trade and trafficking, green criminology, and biodiversity conservation more broadly. It will also be of interest to professionals working in wildlife law enforcement.

Tanya Wyatt is Professor of Criminology in the Department of Social Sciences, Northumbria University, UK. She is the author/editor of numerous books, including *Green Crime in Mexico* (2018), *Green Criminology in the 21st Century* (2016), and *Wildlife Trafficking* (2013).

Routledge Studies in Conservation and the Environment

This series includes a wide range of inter-disciplinary approaches to conservation and the environment, integrating perspectives from both social and natural sciences. Topics include, but are not limited to, development, environmental policy and politics, ecosystem change, natural resources (including land, water, oceans and forests), security, wildlife, protected areas, tourism, human-wildlife conflict, agriculture, economics, law and climate change.

Power in Conservation
Environmental Anthropology Beyond Political Ecology
Carol Carpenter

Rethinking Wilderness and the Wild
Conflict, Conservation and Co-existence
Edited by Robyn Bartel, Marty Branagan, Fiona Utley and Stephen Harris

Humans and Hyenas
Monster or Misunderstood
Keith Somerville

Is CITES Protecting Wildlife?
Assessing Implementation and Compliance
Tanya Wyatt

For more information about this series, please visit: www.routledge.com/Routledge-Studies-in-Conservation-and-the-Environment/book-series/RSICE

IS CITES PROTECTING WILDLIFE?

Assessing Implementation and Compliance

Tanya Wyatt

earthscan
from Routledge

Routledge
Taylor & Francis Group
LONDON AND NEW YORK

First published 2021
by Routledge
2 Park Square, Milton Park, Abingdon, Oxon OX14 4RN

and by Routledge
605 Third Avenue, New York, NY 10158

Routledge is an imprint of the Taylor & Francis Group, an informa business

© 2021 Tanya Wyatt

The right of Tanya Wyatt to be identified as author of this work has been asserted by her in accordance with sections 77 and 78 of the Copyright, Designs and Patents Act 1988.

All rights reserved. No part of this book may be reprinted or reproduced or utilized in any form or by any electronic, mechanical, or other means, now known or hereafter invented, including photocopying and recording, or in any information storage or retrieval system, without permission in writing from the publishers.

Trademark notice: Product or corporate names may be trademarks or registered trademarks, and are used only for identification and explanation without intent to infringe.

British Library Cataloguing-in-Publication Data
A catalogue record for this book is available from the British Library

Library of Congress Cataloging-in-Publication Data
A catalog record for this book has been requested

ISBN: 978-0-367-44071-8 (hbk)
ISBN: 978-0-367-44128-9 (pbk)
ISBN: 978-1-003-00783-8 (ebk)

Typeset in Bembo
by Apex CoVantage, LLC

To Ed, for always supporting me and always being my sounding board for ideas, methods, and rants.

And to Gobi, for being on the couch beside for nearly every second that I was writing this, reminding me why it is important.

CONTENTS

FIGURES

TABLES

BOXES

PREFACE

This book is the latest piece of research after 15 years of being steeped in the world of wildlife trade and wildlife trafficking. It can be a hard topic to be immersed in with the continual decline in species, and the continual coverage of seizures of more dead or suffering wildlife. At the heart of the international trade is the Convention on International Trade in Endangered Species of Wild Fauna and Flora (CITES). There is much to be done to improve this mechanism, which can be achieved if the stakeholders find the will to prioritize wildlife trade and open themselves to having difficult conversations about the need for reform as well as to confront humanity's destructive relationship with the natural world, especially wildlife. I hope that the findings from my research documented in this book can play a part in better protecting wildlife and turning the tide on the biodiversity crisis.

ACKNOWLEDGEMENTS

My heartfelt thanks go out to all the people who took the time to look at and take my survey as well as to the 20 people who were interviewed. Thank you to the UK Arts and Humanities Research Council (AHRC) for the Leadership Fellowship (AH/R002584/1), which allowed me to dedicate 2 years to this in-depth study.

Special thanks to Monica Pons Hernandez for pulling out the data on implementation on my request, which I analyse in Chapter 5.

Special thanks, too, to Alison Hutchinson for helping me to sift through my massive spreadsheet to see the species who were or were not "wildlife".

ABBREVIATIONS

CITES—Convention on International Trade in Endangered Species of Wild Fauna and Flora
IUCN—International Union for the Conservation of Nature
MEA—multinational environmental agreements
NLP—National Legislation Project
NDF—non-detriment finding

1

THE BIODIVERSITY CRISIS

The Earth is facing environmental degradation from many directions. In addition to climate change and plastic pollution, we are also facing a biodiversity crisis. Forests are cut down; the ice sheets are melting; wetlands are converted into farms; deserts takeover grasslands—the list of ways in which biodiversity is being lost goes on. While most environmental damage is driven by converting the land, so it is better suited to human use, the environment, of course, is also seen as a natural resource to be used by people. For many people, wildlife is considered to be a natural resource. Many species of wildlife—animals, plants, fungi, and bacteria (see Nurse and Wyatt 2020 for a full exploration of what wildlife is)—are entangled in the biodiversity crisis. Whereas habitat loss may be the main reason for the loss of biodiversity and species, there is also the loss of species due to direct overexploitation and from illegal trade (CITES 2016a; IUCN 2016; WWF 2017; IPBES 2019), as this book explores.

A biodiversity crisis may sound like a problem to be solved by conservationists, ecologists, and other experts from the natural and life sciences. But conservation and the management of wildlife that stems from those conservation efforts are inextricably linked to law and the legal systems enacting the laws and regulations. Local, regional, national, and/or international administrative, civil, criminal, and environmental regulations are key elements in implementing many conservation strategies. These conservation regulations might take the form of establishing national parks, protected areas, community-based natural resource management schemes, or legal listings of protected species, among other approaches. In regard to overexploitation and illegal trade of wildlife, the main international legal instrument to try to address these is CITES. In May 2018, I began a socio-legal study of CITES funded by the UK Arts and Humanities Research Council. This book captures the overall findings of this leadership fellowship, which sought to contribute to the reduction of overexploitation and illegal trade by identifying best practices and lessons learned by CITES parties in relation to the implementation of and compliance with the Convention.

CITES' purpose is to ensure international trade in wildlife is sustainable. The legislative framework developed and agreed to by the CITES' parties is meant to operationalize this purpose. But, is CITES working and protecting wildlife? A brief exploration of the species CITES is meant to be protecting suggests may be not (and this does not address the glaring issue of all the species not listed on the CITES appendices that may be traded unsustainably). A decline of some traded species is evident from the continued increase in the number of species protected by CITES, which I discuss in more detail in Chapter 2. Further evidence is found in the possible impending extinction of species already listed by CITES such as the rhinoceros, several species of which have fewer than 100 individuals (WWF 2020a), though rhinoceros have been listed in the CITES appendices for decades. Frequently, a lack of enforcement of wildlife regulations is blamed for the decline in protected species (Oldfield 2003; IPBES 2019). Whereas this is likely to be a part of the overall picture, it may be that the regulation regimes and lack of compliance in countries with species decline can (also) benefit from changes to regulation in the form of improved legislation, as well as better implementation and compliance with existing legislation. These improvements may be rather than or in addition to improved enforcement of the existing legislation. Before improved implementation of and compliance with legislation can be explored in depth, it is important to have complete understanding of the state of nature and the role of wildlife trade in the biodiversity crisis. I will then provide a brief introduction to CITES, and why I argue this research is crucial. Afterwards, I will set out the structure for the remainder of the book.

How is wildlife trade contributing to the biodiversity crisis?

According to the IPBES (2019), 75 per cent of the Earth's land surface has been significantly altered, and 66 per cent of the ocean shows signs of cumulative human impacts. This has profound consequences for ecosystems, biodiversity, and species. Of the animal and plant species who have been assessed, approximately 25 per cent of those are threatened, which means, as mentioned, 1 million species may go extinct in the next few decades (IPBES 2019). That is a figure based upon the species who have been assessed; the International Union for the Conservation of Nature (IUCN)—the world's most comprehensive information source for the conservation status of species—has assessed 116,000 of the Earth's species (IUCN 2020). It has been estimated that there are approximately 8.7 million animal, plant, and fungal species globally, but this might be 1.3 million off in either direction (lower or higher) (Mora et al. 2011). About 2.2 million (give or take 0.18 million) are marine species (Mora et al. 2011). These estimates do not include bacteria and archaea. Mora et al. (2011: no page) state:

> In spite of 250 years of taxonomic classification and over 1.2 million species already catalogued in a central database, our results suggest that some 86% of existing species on Earth and 91% of species in the ocean still await description.

The point is that estimates of 1 million species extinctions are based upon incomplete data since assessments of species' conservation status have been completed for only 10 per cent of the species who have been identified. Extinctions may be much higher than predicted.

Whereas IPBES (2019) states the number of extinctions is tens to hundreds of times higher than the extinction rate over the last 10 million years, Wilson (2016) estimates the rate of human-induced species extinction is 100 to 1,000 times higher than non-human-caused extinctions. The main driver of biodiversity loss and species extinction is the high amount of land-use change, which "for terrestrial and freshwater ecosystems, . . . has had the largest relative negative impact on nature since 1970" (IPBES 2019: 12). Land-use change includes, but is not limited to, deforestation, and wetland and grassland conversions, much of which take place for agriculture. The change of land use as the main driver is

> followed by the direct exploitation, in particular overexploitation, of animals, plants and other organisms, mainly via harvesting, logging, hunting and fishing. In marine ecosystems, direct exploitation of organisms (mainly fishing) has had the largest relative impact, followed by land-/sea-use change.
>
> (IPBES 2019: 12)

As Reeve (2002) highlights, in addition to the potential negative effects on species population of collection and harvest from the wild, there are other further possible negative outcomes for the environment and wildlife. For instance, if keystone species are removed, and/or if specific portions of species populations are disproportionately taken, such as more individuals of certain ages, sexes, and physical characteristics, this can have additional damaging impacts on the ecosystems. Furthermore, if certain species of plants are targeted for harvest, for example, this may have adverse effects on the food supply of certain non-human animals and thus, have particularly harmful impacts on those species' populations. An example of this comes from Russia Far East, where deforestation from legal and illegal logging of cedar trees is thought to have contributed to the decline in the wild boar population (Wyatt 2013b). The wild boars' diet largely consists of the seeds and leaves of the cedar. With fewer wild boars, there is less prey to eat for the critically endangered Amur (Siberian) tiger, of which approximately just over 500 remain (BBC 2020; WWF 2020b). This example also raises the issue of deforestation for logging as a form of wildlife trade, since timber is wildlife. Whether or not this is recognized by CITES or the CITES parties is an issue I return to later, but it is important to note that legal and illegal logging can devastate biodiversity as well as contribute to climate change, which has further knock-on effects for species' survival. Finally, Reeve (2002) points out that the methods of collecting and harvesting for trade may also have deleterious impacts on the environment, particularly in cases where destructive techniques are used, such as cyanide and dynamite fishing (which, depending upon the location, may be legal).

A further possible risk to biodiversity, stemming from wildlife trade, is the potential biosecurity threats. These biosecurity threats take two forms: invasive species and disease. In regard to invasive species, this occurs accidentally when wildlife gets shipped around the world either unintentionally or purposely through illegal wildlife trade or when people release non-native species into new ecosystems. Possibly one of the more well-known instances of invasive species is that of the Burmese python in Florida. Burmese pythons, and other large non-native snakes, are imported into Florida for the "exotic" pet trade. These pythons can grow to lengths of nearly 3 metres, which is thought to be the reason why some owners cannot cope with caring for them and thus illegally release them into the environment. The population of Burmese pythons in Florida continues to grow as they outcompete native reptiles and other predators. Harvey et al. (2008: 2) have noted that Burmese pythons have "broad dietary preferences, a long lifespan (15–25 years), a high reproductive output, and [the] ability to move long distances". Furthermore, they have added further pressure on the survival of the local native endangered species, such as the Key Largo woodrat and round-tail muskrat because of the additional predation (Science Daily 2008). Invasive species, then, can play a role in declining levels of biodiversity.

The second biosecurity threat, and of particular contemporary relevance to discussions of wildlife trade, is the link to disease transmission. The ongoing coronavirus pandemic has left no part of the planet untouched, with nearly 70 million infections and over 1.5 million deaths at the time of writing (WHO 2020). The transfer of zoonotic diseases (e.g. virus, bacteria, parasite, and/or fungi from non-human animals to other species, including humans) is a significant threat to human health, with 61 per cent of all human pathogens being zoonotic (WHO 2006). Most recent emerging human infectious diseases have been zoonotic (Swift et al. 2007), and zoonotic diseases have been shown to coincide with the global wildlife trade (Naim 2005). The coronavirus, SARS-CoV-2, likely originated in bats, possibly passing through an unknown second species, and then on to humans through the wildlife trade (Kimbrough 2020), driven, in this case, by human wild meat consumption. COVID-19, the disease resulting from the SARS-CoV-2 virus, is one of many zoonotic diseases likely linked to wildlife trade such as:

- severe acute respiratory syndrome (SARS), from civet cats, who are eaten.
- HIV, Ebola, monkey pox, Hepatitis A and B, Herpes B, shigellosis (a highly infectious form of dysentery), cholera, and tuberculosis from primates as pets and in laboratories that are then sold as pets.

- tapeworms, roundworms, salmonella, and human leprosy from small mammals and reptiles in the pet trade.
- avian flu and psittacosis from birds in the pet trade.

(Green and CPI 1999; Karesh et al. 2005; Swift et al. 2007)

The above are instances of human diseases, but a perhaps overlooked aspect to the biosecurity threat of the wildlife trade is the transmission of diseases between plants and between non-human animals. "Trade in wildlife provides disease transmission mechanisms at levels that not only cause human disease outbreaks but also threaten livestock, international trade, rural livelihoods, native wildlife populations, and the health of ecosystems" (Karesh et al. 2005: no page). As Karesh et al. (2005) note, disease transmission can have significant consequences on biodiversity and species' survival. The global scale of the wildlife trade and of wildlife markets, where species from numerous regions and/or countries are sold, combined with rapid transportation, creates the conditions for naturally occurring diseases to be transmitted between species (Karesh et al. 2005). As stated earlier, this includes disease transmission between plants, which is possibly less visible than between non-human animals. As just one of numerous instances, in the UK, there is the spread of the fungus *Phytophthora ramorum*, a water mould, that was introduced to the UK through the substantial horticulture trade in the country (Wilkinson et al. 2011). This water mould is highly destructive and deadly to over 150 plant species, including some commercial timber species, and can be spread through footwear, dogs' paws, and car and bike tyres, among other ways (Forest Research 2020). The point of this one small example is that trade in plants also poses threats to biodiversity.

In regard to the trade in non-human animals, there are also many instances of disease transmission between non-human animals. For example, in 2002, four green tree pythons arriving in Australia from Singapore were found to be infected with the Wamena virus, a pathogen lethal to a variety of cold-blooded non-human animals such as fish, amphibians, and turtles and tortoises (Hyatt et al. 2002). Fortunately, since the trade was taking place legally, the proper quarantine and health checks were in place, so that the disease was not passed to other non-human animals. As mentioned earlier, in regard to the coronavirus, it is presumed there was an intermediate species between the infected bat and the first human infection. This clearly indicates that disease can be transmitted between diverse species as part of the wildlife trade.

One final point regarding disease transmission, wildlife trade, and biodiversity is the adverse impacts on wildlife when a zoonotic disease passes back from humans to wildlife. The coronavirus has gone from an unknown non-human animal to a human and on to a number of other species. There are reports of tigers, lions, and a puma in zoos around the world being infected with SARS-CoV-2 (in addition to domestic companion animals) (CDC 2020). It is believed that human workers at these facilities passed the coronavirus to the big cats. In laboratory research, "cats, dogs, ferrets, fruit bats, hamsters, and tree shrews can become infected with the virus. Cats, ferrets, fruit bats, and hamsters can also spread the infection to other animals of the same species in laboratory settings" (CDC 2020: no page).

Furthermore, "Rhesus macaques, cynomolgus macaques, Grivets, and common marmosets can become infected SARS-CoV-2 and become sick in a laboratory setting" (CDC 2020: no page). Evidence suggests, however, that at least 60 mammal species could be infected by the coronavirus (Baynes 2020). Thus, humans should take precautions (masks and distance) from captive wildlife and avoid contact with wild populations to not risk transmitting SARS-CoV-2 to already fragile populations, particularly of endangered species (Baynes 2020).

Most notable though is transmission from humans to mink on fur farms. SARS-CoV-2 has been found in mink in nine countries (Denmark, France, Greece, Italy, the Netherlands, Lithuania, Spain, Sweden, and the United States) (OIE 2020). To protect the people, to keep the virus from mutating in the transmissions between humans and mink and thus jeopardizing vaccine effectiveness, and to protect domestic and wild populations of other species in case an infected mink was to escape or come into contact with them, many of these countries have chosen to cull their mink. For instance, Denmark moved to cull all 17 million mink in the country's fur farms (Mallapaty 2020). Similarly, the Netherlands has decided to cull all mink by 2021 and end fur farming (Mallapaty 2020).

Thus, zoonotic disease transmission risks the lives of wildlife. As such, in addition to the threat to biodiversity from the collection and harvesting of wildlife for trade, there is also further risk to biodiversity stemming from possible invasive species being introduced to new ecosystems as well as the spread of disease between wildlife species.

Exploitation from trade stems from both legal and illegal consumption of wildlife. As Robinson and Sinovas (2018) and Scheffers et al. (2019) have pointed out, exploitation of wildlife is one of the most profitable activities in the world. Yet, despite wildlife trade's ubiquity, profitability, and impact, apart from research on the nature and extent of trade in terrestrial vertebrates (Scheffers et al. 2019), there have not been studies of the global patterns of wildlife trade (Fukushima et al. 2020). Fukushima et al. (2020) have attempted to fill this gap by mining the IUCN database of species' conservation status (the Red List). The wildlife who are subject to exploitation are from the entire range of species. The IUCN (2020) itself documents—of the more than 31,000 species, which they have assessed, who are threatened with extinction from all drivers—41 per cent are amphibians, 14 per cent are birds, 34 per cent are coniferous trees, 27 per cent are selected crustaceans, 25 per cent are mammals, 30 per cent are sharks and rays, and 33 per cent are reef corals. In regard to trade, specifically, Fukushima et al. (2020) found they could not draw inferences on the number of non-vertebrates in trade because the IUCN records are biased towards vertebrates; non-vertebrate species and plants tend not to be assessed. That being said, accounting for this bias, they were able to determine,

> that wildlife trading affects a large number of Phyla and Divisions, including plants (Tracheophyta), fungi (Basidiomycota), most major invertebrate groups (Arthopoda, Cnidaria, Mollusca, and Echinodermata), and both terrestrial and aquatic vertebrates (Chordata).
>
> (Fukushima et al. 2020: no page)

Over 70 per cent of critically endangered, endangered, or vulnerable Echinodermata (starfish, sea urchins, and sea cucumbers) and Cnidaria (coral) species are threatened by trade, which is more than twice the number of vertebrates (Fukushima et al. 2020). In their assessment of vertebrates threatened by trade, Scheffers et al. (2019) discovered that groups with particular characteristics, like rare colouring or larger body size, suffer the negative effects of trade most. These observations raise two important points in terms of biodiversity and conservation. First, better efforts need to be made in regard to both the number of species and which species are assessed in order to ensure those species at risk of extinction are identified in a timely manner. This will enable some conservation measures to be put in place. Second, some species are targeted for trade more than others are and, as Scheffers et al. (2019) note, the long-term impact on certain evolutionary lineages is significant. Losing species who fill particular niches in ecosystems can have far-reaching consequences for all biodiversity.

What these studies of the IUCN Red List also bring to the fore is that there are many species outside of CITES, who are being unsustainably traded and may well be endangered or approaching a threshold where their survival may be threatened. Furthermore, CITES species, too, are being overexploited (more on this later) even though CITES is meant to ensure the sustainability of wildlife trade in endangered species and other species in trade listed in the Convention. Unsustainable trade and overexploitation are legal unless specific regulations (like CITES) are breached. While Fukushima et al. (2020) and Scheffers et al. (2019) both mention illegal trade as part of the overall problem, their studies were largely aimed at trade in general and have not drilled down into the legal versus illegal portions. The most comprehensive picture to date of the nature and extent of illegal trade is the United Nations Office on Drugs and Crime's (UNODC 2016, 2020) *World Wildlife Crime Reports*.

UNODC's (2016, 2020) reports are drawn from global CITES seizures. As mentioned earlier, and in the reports, this presents a particular picture of criminality because a significant amount of wildlife is not listed in CITES. The other limitation of this approach is the international focus of CITES, which means that domestic wildlife crimes, like poaching, which might feed the international illegal supply chain, are not captured by CITES data. The data do, however, give some sense of the scale and scope of illegal trade that is part of the biodiversity crisis, though this is limited to seizure data, which is only a fraction of the illegal wildlife trade. In the 2016 report, there were 164,000 seizures in 120 countries in the World WISE database (UNODC 2016). In that report, it is documented that from 1999 to 2015, 30 per cent of the seizures were of mammals, 28 per cent reptiles, 17 per cent corals, 9 per cent birds, 6 per cent bony fish, and 10 per cent other. There were regional trends of these wildlife seizures. For instance, most reptile seizures took place in Asia and most bird seizures in Central and South America (UNODC 2016). The same data limitations exist here as above in the IUCN database, in that CITES lists fewer species from some taxonomic groups than other groups. Thus, the vertebrate bias mentioned earlier likely plays a role in these data as well.

This was recently acknowledged by the current CITES Secretary General—Ivonne Higuero—when she indicated in her first speech to delegates to CITES that there is a need to bring equal attention to lesser-known species (IISD 2019).

In 2020, the updated version of the *World Wildlife Crime Report* contained two further complete years of seizure data. In 2016, the UNODC World WISE database had 17,881 seizures recorded, and in 2018 there were 20,762 seizures (UNODC 2020). These figures are, by far, the most ever recorded in a single year. The next closest figure was 13,492 seizures in 2010 (UNODC 2020). It is not possible to determine the reason for the substantial increase in recorded seizures. It may be a combination of more law enforcement effort, priority, and operations related to wildlife trafficking as well as more of actual wildlife trafficking. With the additional 37,000+ seizures, the proportion of species being traded has changed from the 2016 report, but mammals still make up the largest percentage (23 per cent), followed by reptiles (21.3 per cent) (UNODC 2020).

Whereas there is volatility in wildlife markets, which makes them difficult to predict and impossible to give exact values of the markets, there is evidence that some species may be driven to extinction by the illegal wildlife trade. For instance, though rhinoceros species are listed in Appendix I of CITES (giving them the highest level of protection, with essentially no trade allowed), poaching for horn trafficking is at levels to where the rhino in South Africa is predicted to be extinct by 2040 (Di Minin et al. 2015). In 2008, just over 200 rhinos were poached throughout Africa; this has steadily risen and reached nearly 1,400 in 2015 (UNODC 2016). It appears, fortunately, that 2015 was the peak of rhino poaching, as numbers have decreased since then as has the price of rhino horn (UNODC 2020). There is also the perhaps controversial claim that CITES itself exacerbates or facilitates the illegal wildlife trade. A former trafficker has said that CITES is the key to laundering illegal wildlife (Hruby 2020). Valid CITES documentation (permits) can be obtained to make wild-caught CITES-listed species appear to be captive-bred (Hruby 2020), thus embedding the illegal within the legal.

The significance of illegal wildlife trade in the overall biodiversity crisis has been recognized. Recommendation 34 of the IPBES (2019) report suggests safeguarding of biodiversity by expanding and effectively managing the existing system of terrestrial, freshwater, and marine protected areas. In order to do so, "into the future also entails enhancing monitoring and enforcement systems, managing biodiversity-rich land and sea beyond protected areas, addressing property rights conflicts and protecting environmental legal frameworks against the pressure of powerful interest groups" (IPBES 2019: 41). In particular, the report highlights the need to combat "wildlife and timber trafficking through effective enforcement and ensuring the legality and sustainability of trade in wildlife" (IPBES 2019: 41) beyond protected areas. Furthermore, there is the need for "Better data on the global extent and forms of wildlife trafficking and its impacts on nature and nature's contributions to people" (IPBES 2019: 56). One aspect of data collection mentioned is the "Monitoring of many listed species in the Convention on International Trade in Endangered Species of Wild Fauna and Flora" (IPBES 2019: 55).

As alluded to, one of the main mechanisms of CITES is the listing of species who are at risk because of trade in one of three appendices. In 2011, there were 625 Appendix I animals and 301 Appendix I plants (Wyatt 2013b). For Appendix II-listed species in 2011, there were 4,685 animals and 29,105 plants. There were 629 animals and 301 plants listed in Appendix I in 2016 (CITES 2016a). There were an additional 4,827 animals and 29,592 plants listed in Appendix II (CITES 2016a). In mid-2020, there were 669 Appendix I animals and 334 Appendix I plants (CITES no date c). In regard to Appendix II, there were 4,952 animals and 29,644 plants. It is important to note that there was the statement on the webpage at that time that "The table below shows the approximate numbers of species who are included in the CITES Appendices as of 2 January 2017" and "Please note that these numbers are approximate because there are no agreed lists for some of the higher taxa" (CITES no date c). These disclaimers are important because first, there was a Conference of the Parties (CoP) in August of 2019 in Geneva. Ninety days after the closing of the Conference, more species would have been added to these lists, so the numbers are even higher than the snapshot from 2017. In fact, the update to this webpage reflecting the species numbers, as of 26 November 2019, indicates that there are 687 Appendix I animals and 395 plants; in regard to Appendix II, there are 5,056 animals and 32,364 plants. Second, as the second quote indicates, there are disagreements with the lists, which raises the possibility of uncertainty for the protection of some species and/or loopholes for trading species who should not be traded.

As mentioned earlier, Appendix I species should only be traded in exceptional circumstances (i.e. scientific purposes), with both import and export permits. Appendix II species are allowed to be traded within set quotas and with the proper export permits. The exact processes for permits, quotas, and so forth will be outlined in Chapter 2. The need to include a discussion of the species listed in CITES appendices here is to show that hundreds, sometimes thousands, of species are added to the CITES appendices (particularly Appendix II) every 3 years when the CoP meeting is held. As Wijnstekers (2018: 140)—former Secretary General of CITES—states "Every transfer of a species from Appendix II to Appendix I can therefore be considered as an example of the failure of the Parties to fulfil their obligations under the Convention".

Given the number of species traded, the number of species listed in CITES because trade is posing a threat to their survival, and that the number listed keeps increasing, it is vital to make CITES work. Furthermore, as noted by the IISD (2019: no page), after the 2019 CoP of CITES:

> This CoP highlighted the increasing pressures on CITES as an instrument to counter the rising scale of biodiversity loss, as parties struggled to address pressures other than trade on wildlife populations, including habitat loss, disease outbreaks, and human-wildlife conflict—issues that CITES is not designed to regulate, but must be taken into account when considering what "sustainable use" means.

The scale and scope of wildlife trade also make it clear that it is a significant element in the biodiversity crisis. Improper regulation of wildlife use or lack of enforcement can easily lead to exploitation and unsustainable use (Kakabadse 2011). CITES, then, needs to be improved to protect species, the environment, and biodiversity. It also needs to be improved to protect wildlife as a cultural heritage and as an economic benefit. CITES came about in the early 1970s in response to similar concerns over the loss of biodiversity. How CITES came to be is important background context, which can help inform improvements now.

The Convention on International trade in Endangered Species of Wild Fauna and Flora

The current biodiversity crisis has been sometime in the making. It has been evident for many decades that numerous species and populations of species were being decimated by human activities. Rachel Carson's 1962 *Silent Spring* is an often-cited example of people sounding the alarm that human behaviour and our relationship to nature needed to drastically change to preserve biodiversity and healthy ecosystems. The creation of CITES as a means to stop trade that was contributing to the larger loss of biodiversity was a part of the bigger environmental movement that was gaining moment across the world and across many spheres of environmental degradation.

In the 1950s and 1960s, conservationists around the world were documenting the decline of fauna and flora linked to trade. This is evident since at the Seventh IUCN General Assembly in 1960, governments in attendance were urged "to restrict the import of animals in accordance with the export regulations of the countries of origin" (Wijnstekers 2018: 33). According to Wijnsteker (2018), this was an impossible task, as there was no mechanism for an importing country to know the regulations of an exporting country. Several years later, in 1963, the issue of restricting trade was revisited by the IUCN. At the General Assembly that year, there was a resolution calling for an international convention on regulation of export, transit, and import of rare and threatened wildlife species or their skins and trophies (Reeve 2002; Wijnstekers 2018). Wijnstekers (2018) speculates that this resolution may be the reason why CITES only refers to endangered species in its title when actually its remit is much wider than only regulating trade in endangered species.

In 1964, a draft of the proposed resolution to establish such a convention was circulated at the General Assembly of the IUCN (Wijnstekers 2018). This was revised in 1969 (Reeve 2002) and included a list of species to be protected (Wijnstekers 2018). Further revisions were shared in 1971 (Reeve 2002; Wijnstekers 2018) based upon feedback from 39 governments and 18 non-governmental organizations (Reeve 2002). Then, in 1972, at the United Nations Stockholm Conference on the Human Environment, a resolution was adopted to hold a conference in Washington DC in March of 1973 to further work on a convention regulating wildlife trade (Wijnstekers 2018).

So, after years of negotiations and meetings, 25 articles and four appendices were agreed upon at the "Washington Convention" (Reeve 2002). They came into force for the original ten signatories on 1 July 1975 (Reeve 2002; Wijnstekers 2018).

> CITES establishes the necessary international legal framework for the prevention of trade in endangered species and for an effective regulation of trade in others. It gives producer and consumer countries their share of the joint responsibility and provides the necessary tools for the international cooperation that is so essential for fulfilling this responsibility.
>
> (Wijnstekers 2018: 34)

According to Reeve (2002: 29), "CITES seeks to prevent unsustainable use, not to promote sustainable use over non-use". The key to knowing whether trade is in fact taking place sustainably or not is complete and timely reporting, but this has been (and seems to continue to be) continually lacking (Reeve 2002). It is essential to monitor trade, which is completely dependent upon the reporting criteria and requirements established by CITES (Doc 8.17) (Reeve 2002). As Reeve (2002) notes, failure to submit reports and trade data, submitting documentation late, and/or reports that do not meet the guidelines on what or how to report render the data useless, as do incomplete or inaccurate data. This can lead to incomplete and/or inaccurate understanding of the nature and scope of trade in species, which in turn can lead to incorrect decisions regarding the levels of trade that could be sustainable. At the extreme, such practices could lead to the wrong listing for a species (or failure to list them at all) and, thus, a failure to provide the species with the proper level of protection to prevent unsustainable trade that could ultimately lead to extinction. I will detail the mechanisms within the Convention to address these sorts of non-compliance by the parties at various points.

That was 45 years ago. CITES still has 25 articles and four appendices. That is not to say nothing has changed in the Convention. On the contrary, Wijnstekers (2018: 34–35) notes CITES has been quite flexible and "that the Parties have been able to adapt its instruments to the ever-changing circumstances and new challenges of regulating international wildlife trade. This, however, does not make understanding all the different rules, or even finding them, any easier". Here, I suggest he is referring not only to the extensive text of the Convention itself, but also to the supportive measures (Resolutions and Decisions), where much of the interpretation and guidance to regulation are written. At the time of writing, there are 101 Resolutions in effect (CITES 2020). Resolutions

> are generally intended to provide long-standing guidance over periods of many years. The Resolutions include the guidance provided by the Conference of the Parties on how to interpret the provisions of the Convention but also include: the documents establishing the permanent committees; the budget and work programme of the Secretariat; rules for controlling the

> trade (such as issuing permits and marking specimens in trade); and the texts establishing long-term compliance processes.
>
> (CITES 2020: no page)

In addition to Resolutions, which provide guidance, there are also Decisions. These tend to be actions that are addressed after they are introduced at a CoP, and are deleted after they have been answered/completed. In July 2020, there were 332 Decisions (CITES no date b) that have not been dealt with after their agreement at the CoP in August 2019. There are also 23 Decisions held over from earlier Conferences of the Parties (CITES no date b). For instance, Decision 14.81—Great Whales—sets out that no periodic review of any great whale, including the fin whale, should take place while the International Whaling Commission's moratorium is in effect (CITES no date a). The Decision is from the 14th CoP, which is clear from the numbering of the Decision. The 14th Conference took place in 2007, so Decisions can remain in effect as long as needed. As a point of reference, when Reeve (2002) was exploring CITES implementation, there were 77 Resolutions and 170 Decisions.

What the number of Articles, Resolutions, and Decisions does highlight is the complex and dynamic nature of CITES. The Convention is continually being adjusted and negotiated depending upon the prevailing concerns of the Parties. Some of the concerns have been about the effectiveness of the Convention itself; "At its ninth meeting (Fort Lauderdale 1994), the Conference of the Parties commissioned a review of the Convention's effectiveness" (Wijnsteker 2018: 39). The only outcome appears to be provisions for conducting strategic planning by the Secretariat. The need to scrutinize CITES' effectiveness, particularly in regard to implementation and compliance stems from the hundreds of unexamined Resolutions and Decisions, in addition to the hundreds of proposed additions or changes to listings every 3 years (Reeve 2002).

The state of knowledge and need for this research

CITES is the mechanism to, at least in part, tackle some of the overexploitation and illegal trade that is contributing to the global biodiversity crisis. The Convention is complex and arguably slow, only meeting once every 3 years and wading through hundreds of items of new business and changes. After 45 years in effect, it is possible that the wildlife trade legislation, both related to CITES and across the world at the national level, has improved in ways to better protect wildlife. This, however, does not seem to be the case. Wildlife receives little protection under common law from the threats of habitat loss, overexploitation, and illegal trade, so international conventions, like CITES and the Convention on Biodiversity (CBD), are important means to protect wildlife and habitat (Rodgers 2013). Overexploitation of wildlife is a ubiquitous problem that, as Schalow (2015) argues, the law deliberately and inadvertently avoids discussing. Even though CITES has been adopted by 183 parties (182 countries and the European Union), similar to the general difficulties in transposing global commitments into national legislation, implementation of CITES legislation remains inconsistent, compliance at times lacking, and listed species still face extinction (CITES 2016b; Reeve 2006; Wyatt 2013b). Furthermore, the minimal research that there is related to CITES (Reeve 2002) has largely been limited to individual species and specific countries. Conservationists and political scientists have conducted most of these studies. There has never been an empirical study of the effectiveness of CITES (Reeve 2002), and there is a lack of empirical case studies related to implementation and compliance. A few policy or legal discussions about the pros and cons of CITES have been written (Aguilar 2013; Bowman 2013; Couzens 2013; Sand 2013; Scanlon 2013; Wandesforde-Smith 2016), but these are theoretical and do not draw on concrete examples of approaches to implementation and compliance.

Without national legislation implementing CITES, the regulation of international wildlife trade would be impossible. As Oldfield (2003) has noted, the lack of enforcement is often blamed when wildlife populations decline. Yet, the fault may be with the design and implementation of the regulation (Oldfield 2003), but there has been very little exploration of national implementation of the requirements of CITES (Wandesforde-Smith 2016) or of projects and approaches that might improve it (Bowman 2013). As will be fully detailed in Chapter 2, signatories or parties to CITES are required to adapt their existing national legislation or pass new legislation to meet specific criteria. CITES' National Legislation Project (NLP) monitors parties' legislation. There have not been any studies that I know of investigating the NLP. In addition, there are a range of, often poorly understood, factors which are likely to influence the apparent effectiveness of CITES, such as the soft law mechanisms integral to implementation and compliance (i.e. international action plans, Review of Significant Trade and Non-detriment Findings (NDFs), and so forth) (Aguilar 2013; Munoz 2016). Another of the factors may be the communication and cooperation between national CITES Management and Scientific Authorities (again, see Chapter 2 for the details about CITES), which will affect effective implementation even when the legislation strictly adheres to the required criteria.

Overexploitation of wildlife due to legal and illegal trade is a central aspect of our planet's biodiversity crisis. Wildlife is consumed in the form of food, seafood, medicine, and tonics; furniture; art, décor, and jewellery; fashion; cosmetics and perfumes; and pets, zoos, and breeding (UNODC 2016). All taxa of wildlife are subject to trade though, as mentioned, most research has only focused on (terrestrial) vertebrates (Fukushima et al. 2020; Scheffers et al. 2019). Even though the research is limited, the data that we do have show that tens of thousands of species of wildlife are likely to go extinct if we do not do something to curb their overexploitation and prevent them from being trafficked. In addition, wildlife trade further threatens biodiversity by being a means of introducing invasive species to new ecosystems and of spreading diseases between wildlife species. Both of these biosecurity risks can result in further loss of biodiversity and can originate in both the legal and illegal trade of wildlife (Wyatt 2013a). Internationally, CITES is the "cornerstone multilateral environmental agreement lying at the nexus of international trade and species conservation" (Reeve 2002: 5). It is *the* mechanism to curb overexploitation and trafficking, and thus a study of how CITES parties are implementing it and complying with it is an important undertaking to help stop the biodiversity crisis.

The layout of the book

Chapter 2 provides the details of how CITES works (including how it is funded), how it has evolved since it was ratified in 1975, and the ongoing limitations and tensions such as protectionism versus utilization as well as at-risk species with low population numbers not getting listed. This is essential to understand how implementation of and compliance with the Convention are falling short of expectations. This is followed by Chapter 3 detailing the mixed methods that I used to collect empirical original data. As I mentioned, according to Reeve (2002), there has never been an empirical evaluation of the implementation of CITES. This book documents the primary research undertaken to fill this gap. Chapter 3 details the methods for the legislative analysis conducted, the Delphi iterative survey undertaken, and the semi-structured interviews done, which, in combination, led to the three case studies of lessons learned as well as an overview of the implementation of and compliance with CITES by all of its members. The justification for the mixed methods is laid out as well as the limitations of such an approach.

Chapter 4—"CITES Implementation: the big picture"—is the first findings chapter. This chapter details the results of the legislative content analysis of the 183 parties' wildlife trade and relevant legislation. Whereas CITES is touted as one of the most successful conventions, over half of the parties still have implementation gaps in their legislation. This chapter breaks down the nature and extent of the member countries' failure to implement the four main components: designate one Scientific and one Management authority, prohibit trade in violation of the Convention, penalize trade violations, and provide for confiscation of illegally trade and possessed wildlife. I also share what experts and CITES reports identify as the main limitations and challenges to implementing CITES. Even though there are only four components by which CITES parties are assessed, there are additional requirements of the Convention. Therefore, this chapter also explores how parties monitor captive breeding, how parties approach their permit system, and how parties have implemented animal welfare elements of CITES into their legislation.

"CITES Compliance: the big picture" is the focus of Chapter 5. This chapter analyses the ways in which parties have and have not complied with CITES requirements. For instance, I detail the quantity and quality of engagement with the annual and biennial reports that parties send to the CITES Secretariat. I also explore how compliance is compromised through lack of enforcement by the Secretariat of the CITES' provisions. In Chapter 6—The lessons learned: highlights from the three case studies—I detail how the case studies were identified and chosen. I offer the thoughts of the experts from the semi-structured interviews related to each of the case studies, and I highlight the best practice and lessons learned overall.

The last chapter—The Future—begins with recommendations for improvements to CITES implementation and compliance. I go on to speculate whether CITES is still fit for purpose or whether a new convention or approach needs to be taken in order to stop the decline of species. Former CITES Secretary General John Scanlon suggests that CITES should incorporate public health elements to address the biosecurity risks, particularly disease transmission, which I mentioned early. He also proposes that a completely new convention tackling wildlife crime in general may be needed, or the global community should consider including wildlife crime and trafficking within existing structures, such as the UN Convention on Transnational Organized Crime (UNTOC). I will explore these options through the use of primary data, asking experts this very question and providing my own analysis of the advantages and disadvantages of these suggestions in regard to wildlife crime and trafficking compared to the current functioning of CITES. The final chapter ends with a discussion of my "Unanswered concerns". There are numerous tensions between CITES parties that, unless tackled head on, will continue to impede efforts to improve the Convention and thus better protect wildlife who are traded and trafficked.

Summary and conclusion

The overexploitation of wildlife, including the illegal wildlife trade as well as the related biosecurity risks, is one of many threats to the rich diversity of life on our planet. This threat, combined with habitat loss, pollution, and climate change, means that 1 million species are facing extinction. It is urgent that efforts are made to address the plethora of threats, and one particular area for improvement is the regulation of international wildlife trade through the Convention on International Trade in Endangered Species of Wild Fauna and Flora. In the words of Ivonne Higuero—the current CITES Secretary General—"Humanity needs to respond to the growing extinction crisis by transforming the way we manage the world's wild animals and plants. Business as usual is no longer an option" (IISD 2019: 57). This book is one (may be the only) empirical attempt to identify some improvements that are critical to stem the loss of biodiversity and species from global trade. To fully understand how CITES can be implemented and complied with better, the next chapter details how CITES currently works.

References

Aguilar, S. 2013. Regulatory tools for the management of fish and timber species through CITES. *Review of European Community & International Environmental Law*. 22 (3): 281–290.

Baynes, C. 2020. Humans told to socially distance around wild animals to avoid 'considerable risk' of passing on coronavirus. *The Independent*. Available at: www.independent.

co.uk/news/science/coronavirus-wild-animal-transmission-covid-19-social-distancing-b864357.html. Accessed 10 December 2020.

BBC. 2020. Endangered tigers making a 'remarkable' comeback. Available at: www.bbc.co.uk/news/newsbeat-53581028. Accessed 10 December 2020.

Bowman, M. 2013. A tale of two CITES: Divergent perspectives upon the effectiveness of the wildlife trade convention. *Review of European Community & International Environmental Law*. 22 (3): 228–238.

Carson, R. 1962. *Silent Spring*. New York: Penguin Books.

Centers for Disease Control and Prevention. 2020. COVID-19 and animals. Available at: www.cdc.gov/coronavirus/2019-ncov/daily-life-coping/animals.html. Accessed 10 December 2020.

Convention on the International Trade in Endangered Species of Wild Fauna and Flora (CITES). No date a. 14.81 great whales. Available at: www.cites.org/eng/dec/valid17/220.7 July 2020.

———. No date b. Decisions of the conference of the parties to CITES in effect after the 18th meeting. Available at: www.cites.org/eng/dec/index.php. 7 July 2020.

———. No date c. The CITES species. Available at: www.cites.org/eng/disc/species.php. Accessed 3 July 2020.

———. 2020. Resolutions of the conference of the parties in effect after the 18th meeting. Available at: www.cites.org/eng/res/index.php. Accessed 7 July 2020.

———. 2016a. The CITES species. Available at: https://cites.org/eng/disc/species.php. Accessed 26 July 2016.

———. 2016b. National laws for implementation of the convention. Available at: https://cites.org/sites/default/files/eng/com/sc/66/Inf/E-SC66-Inf-19.pdf. Accessed 26 July 2016.

Couzens, E. 2013. CITES at forty: Never too late to make lifestyle changes. *Review of European Community & International Environmental Law*. 22 (3): 311–323.

Di Minin, E., Laitila, J., Montesino-Pouzols, F., Leader-Williams, N., Slotow, R., Goodman, P. S. and Moilanen, A. 2015. Identification of policies for a sustainable legal trade in rhinoceros horn based on population projection and socioeconomic models. *Conservation Biology*. 29 (2): 545–555.

Forest Research. 2020. Ramorum disease (*Phytophthora ramorum*). Available at: www.forestresearch.gov.uk/tools-and-resources/pest-and-disease-resources/ramorum-disease-phytophthora-ramorum/. Accessed 10 December 2020.

Fukushima, C., Mammola, S. and Cardoso, P. 2020. Global wildlife trade permeates the tree of life. *Biological Conservation*. 247: 180503. https://doi.org/10.1016/j.biocon.2020.108503.

Green, A. and the Center for Public Integrity (CPI). 1999. *Animal Underworld: Inside America's Black Market for Rare and Exotic Species*. New York: Public Affairs.

Harvey, R., Brien, M. L. Cherkiss, M. S., Dorcas, M., Rochford, M., Snow R. W. and Mazzotti, F. J. 2008. Burmese pythons in South Florida: Scientific support for invasive species management. Available at: http://edis.ifas.ufl.edu/pdffiles/UW/UW28600.pdf. Accessed 28 May 2009.

Hruby, D. 2020. He once trafficked in rare birds. Now, he tells how it's done. *The New York Times*. Available at: www.nytimes.com/2020/11/27/world/europe/rare-birds-macaws-trafficking-brazil.html. Accessed 1 December 2020.

Hyatt, A., Williamson, M., Coupar, B., Middleton, D., Hengstberger, S. and Gould, A. 2002. First identification of a ranavirus from green pythons (*Chondropython viridis*). *Journal of Wildlife Diseases*. 38: 239–252.

Intergovernmental Science-Policy Platform on Biodiversity and Ecosystem Services (IPBES). 2019. IPBES global assessment summary for policymakers. Available at: www.ipbes.net/sites/default/files/downloads/summary_for_policymakers_ipbes_global_assessment.pdf. Accessed 7 May 2019.

International Institute for Sustainable Development (IISD). 2019. Earth negotiations bulletin. August. Available at: http://enb.iisd.org/vol21/enb21101e.html?utm_medium=email&utm_campaign=ENB%20Update%20%2030%20August%202019%20SW&utm_content=ENB%20Update%20%2030%20August%202019%20SW+CID_a7fade43500fa9f1ca5d10a7a4d2b3fa&utm_source=cm&utm_term=Read. Accessed 21 October2019.

International Union for the Conservation of Nature (IUCN). 2020. Homepage. Available at: www.iucnredlist.org. Accessed 3 July 2020.

———. 2016. Categories and Criteria. Available at: http://s3.amazonaws.com/iucnredlist-newcms/staging/public/attachments/3097/redlist_cats_crit_en.pdf. Accessed 26 July 2016.

Kakabadse, Y. 2011. The fight against wildlife crime—enforcement v. corruption. *Environmental Policy and Law*. 41 (3): 123–126.

Karesh, W., Cook, R., Bennett E. and Newcomb, J. 2005. Emerging infectious diseases. Available at: www.cdc.gov/ncidod/EID/vol11no07/05-0194.htm. Accessed 20 May 2011.

Kimbrough, L. 2020. Conservationists set the record straight on COVID-19's wildlife links. *Mongabay*. Available at: https://news.mongabay.com/2020/03/conservationists-set-the-record-straight-on-covid-19s-wildlife-links/. Accessed 16 March 2020.

Mallapaty, S. 2020. COVID mink analysis shows mutations are not dangerous—yet. *Nature*. 587: 340–341.

Mora, C., Tittensor, D. P., Adl, S., Simpson, A. G. B. and Worm, B. 2011. How many species are there on earth and in the ocean? *PLoS Biology*. 9 (8): e1001127. https://doi.org/10.1371/journal.pbio.1001127.

Munoz, M. A. C. 2016. The role of CITES in ensuring sustainable and legal trade in wild fauna and flora. In Elliott, L. and Schaedla, W. (eds.). *Handbook of Transnational Environmental Crime*. Cheltenham: Edward Edgar: 433–442.

Naim, M. 2005. *Illicit: How Smugglers, Traffickers and Copycats Are Hijacking the Global Economy*. London: William Heinemann.

Nurse, A. and Wyatt, T. 2020. *Wildlife Criminology*. Bristol: Policy Press.

Oldfield, S. 2003. Introduction. In Oldfield, S. (ed.). *Trade in Wildlife: regulation for Conservation*. London: Earthscan: xvii–xxvii.

Reeve, R. 2006. Wildlife trade, sanctions and lessons from the CITES regime. *International Affairs*. 82: 881–897.

———. 2002. *Policing International Trade in Endangered Species: The CITES Treaty and Compliance*. London: Earthscan.

Robinson, J. E. and Sinovas, P. 2018. Challenges of analyzing the global trade in CITES-listed wildlife. *Conservation Biology*. 32 (5), 1203–1206. https://doi.org/10.1111/cobi.13095.

Rodgers, C. 2013. *The Law of Nature Conservation*. Oxford: Oxford University Press.

Sand, P. 2013. Enforcing CITES: The rise and fall of trade sanctions. *Review of European Community & International Environmental Law*. 22 (3): 251–263.

Scanlon, J. 2013. CITES at its best: CoP16 as a 'watershed moment' for the world's wildlife. *Review of European Community & International Environmental Law*. 22 (3): 222–227.

Schalow, F. 2015. Heidegger, the law of being and animal protection laws. *Ethics & the Environment*. 20 (2): 61–82.

Scheffers, B., Oliveira, B., Lamb, I. and Edwards, D. 2019. Global wildlife trade across the tree of life. *Science*. 366 (6461): 71–76. https://doi.org/10.1126/science.aav5327.

Science Daily. 2008. Python snakes, an invasive species in Florida, could spread to one third of us. Available at: www.sciencedaily.com/ releases/2008/02/080223111456.htm. Accessed 28 May 2009.

Swift, L., Hunter, P., Lees, A. and Bell, D. 2007. Wildlife trade and the emergence of infectious diseases. *EcoHealth*. 4 (25): 25–30.

United Nations Office on Drugs and Crime (UNODC). 2020. *World Wildlife Crime Report: Trafficking in Protected Species*. Vienna: UNODC.

———. 2016. *World Wildlife Crime Report: Trafficking in Protected Species*. Vienna: UNODC.

Wandesforde-Smith, G. 2016. Looking for law in all the wrong places? Dying elephants, evolving treaties & empty threats. *Journal of International Wildlife Law & Policy*. 19 (4): 365–381.

Wijnstekers, W. 2018. *The Evolution of CITES: A Reference to the Convention on International Trade in Endangered Species of Wild Fauna and Flora, 11th Edition*. International Council for Game and Wildlife Conservation. Available at: www.cic-wildlife.org/wp-content/uploads/2018/05/the-evolution-of-cites-2018-11th-editionpart-1.pdf. Accessed 7 July 2020.

Wilkinson, K., Grant, W., Green, L., Hunter, S., Jeger, M., Lowe, P., Medley, G., Mills, P., Phillipson, J., Poppy, G. and Waage, J. 2011. Infectious disease of animals and plants: an interdisciplinary approach. *Philosophical Transactions of the Royal Society of London. Series B, Biological Sciences*. 366 (1573): 1933–1942.

Wilson, E. O. 2016. *Half Earth: Our Planet's Fight for Life*. New York: Liveright Publishing.

World Health Organisation (WHO). 2020. Coronavirus disease (COVID-19) pandemic. Available at: www.who.int/emergencies/diseases/novel-coronavirus-2019. Accessed 10 December 2020.

———. 2006. Neglected zoonotic diseases. Available at: www.who.int/neglected_diseases/diseases/zoonoses/en/. Accessed 01 January 2020.

World Organisation for Animal Health (OIE). 2020. COVID portal—events in animals. Available at: www.oie.int/en/scientific-expertise/specific-information-and-recommendations/questions-and-answers-on-2019novel-coronavirus/events-in-animals/. Accessed 10 December 2020.

World Wildlife Fund (WWF). 2020a. Rhino. Available at: www.worldwildlife.org/species/rhino. Accessed 22 June 2020.

———. 2020b. Tiger facts. Available at: www.worldwildlife.org/species/tiger. Accessed 10 December 2020.

———. 2017. Unsustainable and illegal wildlife trade. Available at: http://wwf.panda.org/about_our_earth/species/problems/illegal_trade/. Accessed 01 February 2017.

Wyatt, T. 2013a. The security implications of the illegal wildlife trade. *Journal of Social Criminology: Special Issue on Green Criminology*. Autumn: 130–158.

———. 2013b. *Wildlife Trafficking: A Deconstruction of the Crime, the Victims and the Offenders*. London: Palgrave Macmillan.

2

HOW DOES CITES WORK?

As outlined in Chapter 1 on the biodiversity crisis, CITES is a complex, slow-moving convention, which—considering that there are 183 parties—has hundreds of organizations and thousands of individuals involved in its functioning. Chapter 2 provides the details of what exactly is meant by implementation and compliance, how CITES works—in particular, the NLP is outlined, along with an overview of the compliance process, the CITES Secretariat and committees as well as how CITES is funded—the ongoing limitations and tensions within CITES, such as protectionism of species versus utilization and sustainable use, and contentious political debates over which species are listed in the appendices. A clarification of the vocabulary as well as the functioning of the Convention is essential to understand how implementation of and compliance with the Convention are or are not falling short of expectations. As Reeve (2002: 15) states,

> an uncontested fact is that uncontrolled international trade has led to drastic depletion of some wildlife species. . . . Ensuring that CITES achieves its objective of preventing international trade from threatening the survival of species is in the interest of all . . . and this is where the compliance system comes in.

Definitions

As Mitchell (1996: 17) notes "There is no consensus as to the meaning of compliance, implementation and enforcement. The concepts overlap, the terms are used interchangeably, and definitions vary". I conceptualize the three elements along a timeline—implementation, compliance, enforcement. A convention, like CITES, is written, and the first step is that parties agree to it and become signatories to that convention. By signing the convention, the parties promise to implement the text of the convention. In the case of CITES, that means to transpose the requirements of the Convention into the parties' national or supranational legislation. I mention supranational legislation because the European Union (EU) is a party to CITES, so EU legislation transposes the requirements of CITES, and the 27 member countries further transpose the EU Directive(s) into their national legislation. Implementation, in this book, specifically refers to this step, where the parties make their national legislation adhere to CITES. This is in line with stricter definitions that define implementation as the measures parties take to make international agreements operate in their domestic law (Finnish Ministry of the Environment 1995). Other definitions are wider and conceptualize implementation to not only be the adoption of national measures implementing an international convention, but also include that those measures are being complied with by the parties and that the measures are being complied with by the entities who are subject to the national measures (Reeve 2002). I find this terminology leads to lack of clarity and possibly confusion as to what is implementation and what is compliance. Due to this, I have chosen the stricter definition to be as clear as possible as to what I am referring to when I say implementation.

After a party has implemented a convention, they should comply with the provisions set out in the convention. "*Compliance* is generally understood in international law as 'behaviour that conforms to a treaty's explicit rules'" (Mitchell 1996: 17, emphasis in original). CITES compliance then is confined to the parties meeting the reporting requirements, meeting the timelines for implementing changes and/or additions, and undertaking the required assessments and monitoring. For example, CITES parties are required to submit annual and biennial reports to the CITES Secretariat. These reports are due in October of the following year (for example, in regard to an annual report in October 2021 for the year 2020) to enable parties enough time to collate their trade and permit data and to gather information related to implementation asked about in the biennial reports. Thus, to be compliant, parties need to submit the reports and submit them on time. As discussed in Chapter 5, a minority of CITES parties are compliant. Therefore, in this book, compliance is a reference to the *parties'* actions, rather than those individuals and organizations who are subject to CITES trade regulations. Although a person or an organization not following the regulations may also be termed "non-compliant", I am making the distinction that these are *breaches* of CITES.

After a convention has been implemented, the parties are required to comply with detailed provisions, and one of those provisions may be to enforce the convention. Enforcement has been defined as the formal legally circumscribed reaction to a breach of an obligation (Mitchell 1996). Again, the breach is on the part of the individuals and organizations subject to the convention. Mitchell (1996) suggests that a distinction can be drawn between national and international enforcement. He states that national enforcement is the reaction to breaches or violations of national/domestic law, most likely by individuals and organizations. International law enforcement, then, is when organizations try to hold states to their international legal obligations. An example would be a country bringing to the attention of the United Nations that another country was violating its citizens' human rights. In the case of CITES, this is possible that a party may bring to the attention of the Secretariat that another party is not fulfilling its obligations under CITES. The CITES Secretariat, too, may engage in "enforcement" activities if parties are not complying with the requirements of the Convention. For the most part, when discussing enforcement, I am referring to parties' efforts to detect breaches of CITES by individuals and organizations. I will, however, discuss in Chapter 5, CITES' efforts to get parties to comply with their obligations, which could be conceptualized as a form of enforcement, but, as Reeve (2002) notes, might better be thought of as managing the Convention. It is important to point out, as Mitchell (1996) rightly does, that compliance by parties to their treaty/convention obligations does not mean that the implementation or the convention is effective. What is known about the state of CITES implementation and compliance as well as the specific required components that make up CITES is now detailed.

CITES implementation requirements and the NLP

Implementation of CITES

Even though CITES is recognized as a significant achievement, it is not without its flaws or without critics. Implementation of CITES into national legislation has received attention over the years as to whether or not parties are implementing the control measures of the Convention. For instance, Yeater (2002) pointed out, at a workshop in 2001, a number of areas that could be improved. The weaknesses indicated were related to not only implementation of the legislation required, but also implementation of the control measures required by the Convention. In regard to the implementation of the legislation, Yeater (2002) identified the lack of or insufficient national legislation, particularly regarding penalties.

In terms of control measures, in order to trade in CITES-listed species, the Convention sets out a permit system regulating the trade of wildlife attached to the listing of species in the appendices, as I introduced in Chapter 1. Appendix I species (those only traded in exceptional circumstances) need to have an export and an import permit; Appendix II species (trade volumes have set limits) need to have an export permit. Permits should not just be issued every time an application is received. The Management and Scientific Authorities (which I detail in a moment) should coordinate so that trade will not threaten the survival of the species. Implementing the permit system seems to be fraught with difficulties. It has been observed that sometimes there are issuance of irregular documents; a lack of or insufficient coordination and communication between the Management authority, Scientific authority, and enforcement agencies; and permit fraud (Yeater 2002). I discuss the approach to permits by different parties in Chapter 4. In addition, there is insufficient communication with the CITES Secretariat in Geneva (Yeater 2002). Other weaknesses that have been highlighted are those related to enforcing the Convention, such as the lack of or insufficient border control and the lack of or insufficient control of domestic trade, since domestic trade has implications for international trade (Yeater 2002).

Around the same time as the workshop cited earlier, where Yeater (2002) detailed concerns with CITES' implementation and enforcement, Reeve (2002) wrote her comprehensive book *Policing International Trade in Endangered Species: The CITES Treaty and Compliance*. This appears to be the only thorough assessment to date of CITES' implementation and compliance. In it, among other points, she lays out the observed weaknesses and a list of what the CITES Secretariat proposed good implementation would look like (implementation for CITES is more than transposing the Convention into national legislation):

> 1 adequate national legislation to regulate CITES trade;
> 2 adequate national legislation to protect and regulate the harvest of species of conservation concern;
> 3 adequate national legislation to enable enforcement and penalize offenders;
> 4 economic incentive policies, incorporated into legislation where necessary, to promote compliance;
> 5 sufficient provision and training of administrative and enforcement personnel;
> 6 provision of effective scientific advice for both administrative and enforcement personnel;
> 7 trade monitoring and analyses combined with information management systems, to aid policy-making;
> 8 education and awareness-raising campaign directed towards traders and the public;
> 9 support of the judiciary and adequately responding to crimes and helping to deter offenders;
> 10 inter-agency cooperation and exchange of information at national, regional and international levels.
>
> (CITES 2001)

The CITES Secretariat has, for many years, undertaken initiatives to improve implementation and compliance. The main initiative, the NLP, started in 1992, which I will return to in a moment. The latest effort may be that the required biennial reports have become implementation reports (Wijnstekers 2018) that break down the parties' efforts to implement the Convention, including the challenges and limitations that they face (see Chapter 5 for more detail). In regard to when biennial reports should be submitted to the Secretariat:

> [the Secretariat] URGES all Parties to submit their reports required under the provisions of Article VIII, paragraph 7 (b), one year before each meeting of the Conference of the Parties, beginning with the 17th meeting of the Conference of the Parties [2016] and in accordance with the report format

> distributed by the Secretariat, as may be amended by the Secretariat from time to time with the concurrence of the Standing Committee.
>
> (in Wijnstekers 2018: 347)

Ironically perhaps,

> Until CoP16 [Conference of the Parties 16, in 2013], the biennial reports were, like the annual reports, to be submitted by 31 October following the year for which they are due. The above new timing makes that the biennial reports are to be submitted only every three years, which is not in accordance with the Convention.
>
> (Wijnstekers 2018: 347)

Therefore, the new timelines for reporting do not meet the requirements set out in the Convention.

Reeve (2002) largely seemed to agree with CITES' own assessment of what was needed for good implementation and found that many elements were lacking. The NLP, which, as mentioned, started 10 years prior to Reeve's study, was and is particularly focused on assessing the implementation into a parties' national legislation of the four main components of CITES.

The National Legislation Project

The CITES' NLP is the initiative of the Convention started in 1992 that oversees categorizations of the CITES parties' national legislation. The NLP is designed to analyse and assist members in drafting and implementing legislation that will meet the CITES standards. The CITES Secretariat categorizes national legislation of parties into three groups (CITES 2016b). Category 1 indicates that the legislation is generally believed to meet the requirements of implementation. For parties' national legislation in Category 2, their legislation is believed generally not to meet all of the requirements for implementation. Category 3 indicates the legislation is believed generally not to meet the requirements for implementation.

When parties' national legislation was first assessed between 1992 and 1994, of the 81 countries that were assessed, only 15 parties (19 per cent) met the requirements to be in Category 1 (Reeve 2002: 134). A further 27 parties (33 per cent) did not meet any of the requirements, so were placed in Category 3. The rest of the parties were judged as Category 2, which means their legislation did not meet all of the requirements (52 per cent) (Reeve 2002: 134–135). Reeve (2002) charts the progress of parties' implementation of the CITES legislation by tracking the categorizations issued through the NLP. In 1997, an additional 44 parties' legislation was analysed. Nine parties' legislation fell into Category 1 (20 per cent), 15 into Category 2 (34 per cent), and 20 (45 per cent) into Category 3. Furthermore, there was some improvement to the parties' legislation of the original 81 assessed, but a majority of the legislation had not improved (Reeve 2002).

In 2000, Reeve (2002) notes further improvement to 37 parties' national legislation being Category 1 (26 per cent), 52 Category 2 (36 per cent), and 47 Category 3 (32 per cent). After the assessment in 2000, further efforts were initiated to improve implementation. In particular, during regional workshops, models of legislation were developed (Reeve 2002). The idea was to try to harmonize legislation across regions as well as to harmonize implementation procedures (Reeve 2002). CITES parties can face sanctions for failure to implement the requirements of the Convention; these compliance measures will be detailed in the section coming up on compliance. Despite CITES having "the stick" of sanctions, initially the NLP found that 75 per cent of parties' legislation had not been adequately implemented (Reeve 2002: 147). As evident from above, this had barely improved by 2000, when inadequate legislation was still found for 68 per cent of the parties' legislation. Implementation had improved by 2002, when 51 per cent of parties' legislation fell in Categories 2 and 3 (Reeve 2002). Since then, there have been many more signatories. Unfortunately, however, adequate implementation is still lacking from the legislation of half of the parties. Currently, in Category 1 there are 92 parties' legislation or just over 50 per cent. Category 2 has 46 parties' legislation (25.3 per cent) and Category 3 has 36 parties' legislation (19.8 per cent); eight parties have recently joined, and their legislation has not been assessed.

Further implementation support has been through the CITES Secretariat's production of a Model Law and checklist for implementation. These are, however, general guidelines and not mandatory processes. It is clear that full implementation is still lacking, and in the next section I detail what the four main components of implementation assessed by the NLP are, which have not changed since the first assessments (Reeve 2002).

The four components of the NLP

The NLP assesses parties' implementation of CITES through four components: designate at least one Management Authority and one Scientific Authority; prohibit trade in specimens in violation of CITES; penalize such trade; and confiscate specimens illegally traded or possessed. I will describe each of the components later.

Authorities

Article IX of the CITES Convention stipulates that each party must designate one or more Management and Scientific Authorities. The two authorities have distinct but intertwined responsibilities and roles. The Management Authority is supposed to be "competent" to issue permits and certificates for the trade of CITES-listed species (CITES 1973). It is debatable what would constitute competency in this case, but presumably the authority must have the infrastructure and technological capabilities to accept applications for permits, process the applications, and issue a permit. Perhaps surprisingly in the modern digital age, permit applications and particularly the actual export and import permits remain paper-based. And, as Reeve (2002) notes, are far less relevant and effective in a world with far fewer borders (i.e. the European Union). The Management Authority issuing export permits (required for Appendix I and II species) must be satisfied that the "specimen" being exported has been obtained legally (CITES 1973). Furthermore, it is the duty of the Management Authority that any live wildlife is "prepared and shipped as to minimize the risk of injury, damage to health or cruel treatment" (CITES 1973: no page). The Convention text makes it clear that there can be one or several of these authorities.

The responsibility for the number of permits allowed and, at least in part, the recommendations for the listing of species in the CITES appendices falls to the Scientific Authority. According to Raymakers (2002), the Scientific Authority should take into account not only the biological characteristics of the species in question, but also the impacts of trade in that species on the place from where they were collected, the likely survival or mortality rate of that species prior to being exported, the level to which that species is consumed domestically for subsistence and commercially, and other threats, such as habitat destruction and pollution. Today, there is also the need for the Scientific Authority to consider the added threat of climate change. Ultimately, the Scientific Authority is responsible for determining that import and/or export "will not be detrimental to the survival of that species" (CITES 1973: no page). As Reeve noted in 2002, NDFs are not specified in the Convention or in the Resolutions. NDFs, though, have become a crucial aspect of CITES trade in the attempt to ensure trade is sustainable. As with the Management Authority, parties are allowed to designate more than one Scientific Authority and, as I will detail in Chapter 4, parties have a variety of ways of structuring their implementation of the Convention.

Prohibition

Article VIII of CITES opens saying, "The Parties shall take appropriate measures to enforce the provisions of the present Convention and to prohibit trade in specimens in violation thereof" (CITES 1973: no page). The Convention does not prescribe how this needs to be done. This means there is no obligation for parties to prohibit violations through any specific legal regime—administrative, civil, criminal, environmental, trade, and/or wildlife. Therefore, there is a tremendous amount of flexibility and discretion as to how CITES parties will prohibit trade that is counter to the Convention provisions (see Chapter 4).

CITES is clear in Article XIV that parties can prohibit more than just violations of CITES, so they may institute:

> (a) stricter domestic measures regarding the conditions for trade, taking, possession or transport of specimens of species included in Appendices I, II and III, or the complete prohibition thereof; or

(b) domestic measures restricting or prohibiting trade, taking, possession or transport of species not included in Appendix I, II or III.

(CITES 1973: no page)

When discussing best practices and findings in Chapter 6, I will return to domestic measures.

Penalization

As with prohibition, CITES does not stipulate how violations of the Convention should or need to be penalized. Article VIII, "Measures to be Taken by the Parties", simply states that measures to enforce and prohibit the Convention shall include "to penalize trade in, or possession of, such specimens, or both" (CITES 1973: no page). Again, this creates a large amount of leeway for how parties implement the penalization component. This may entail fines, prison, or a combination as well as alternative approaches, as I will discuss in Chapter 4.

Confiscation

Article VIII, requiring prohibition including penalizing violations, also mandates parties "to provide for the confiscation or return to the State of export of such specimens" (CITES 1973: no page). The NLP checks that confiscation is included in the implementation of CITES into national legislation, but it does not appear to go further than that. In addition to confiscation of both wildlife products and live wildlife, it would seem important to check whether parties securely and safely store wildlife products that they have confiscated, so that they can be used as evidence for issuing penalties and to check whether live wildlife who are confiscated is returned to the export country or housed in a rescue centre or other suitable facilities under the proper conditions, as stipulated in Article VIII.

In Chapter 3, I will outline how I went about analysing whether these four components were or were not present in the parties' legislation. Now, I expand upon what is known about parties' compliance with the requirements of CITES.

Compliance

The previous section showed what the main components of implementation of CITES legislation should be and that even though CITES is 45 years old, around half of the parties have still only partly implemented the legislation or have not implemented it at all. Unfortunately, the parties' track record for compliance with other requirements is also not adequate. As mentioned, the Convention sets out annual reporting of trade volumes and biennial reporting of implementation to the CITES Secretariat by 31 October in the following calendar year. Meeting these deadlines and reporting requirements are monitored by the CITES Secretariat and the Standing Committee (which I detail shortly). Unlike other multinational environmental agreements, CITES has the ability to use trade restrictions as a sanction or a "stick" to try to prompt parties to implement the Convention and to comply with reporting requirements (Reeve 2002). Reeve (2002) claims that the use of the stick on seven parties in the late 1990s for failure to implement CITES legislation in the national context resulted in improvement of all those targeted. However, she notes, application of sanctions must be equally administered, which was not the case in the early 2000s. Further, she warned that the fear of trade sanctions was short-lived as it was not used regularly and ceased to be a deterrent.

Despite having the ability to issue trade sanctions for failure to comply with implementation and reporting requirements, it is not done all that often. Annual reporting is largely ignored by parties and NGOs alike, and, at least up until 2002, the biennial reporting requirement was largely unimplemented, and there was little follow-up (Reeve 2002). As I will detail in Chapter 5, these trends remain largely true, with meeting the annual and biennial reporting requirements being the exception rather than the norm. That is not to say that the CITES Secretariat and the Standing Committee take no action. Currently, as I discuss in detail in Chapter 5, there are a number of parties listed as in need of urgent action regarding implementation and/or reporting; there are over 30 parties under some form of trade suspensions.

Sanctions seem to be a measure of last resort. There are many steps taken before the stick is used. The way to deal with non-compliance has evolved through the Resolutions and Decisions of CITES rather than being outlined in the Convention text. Article XIII of the Convention mentions two elements of concern—trade that is adversely affecting a species and Convention provisions not being effectively implemented (CITES 1973). The only procedure mentioned in the Convention text is the Secretariat contacting the relevant Management Authorities and those authorities deciding whether or not an inquiry needs to be made. The results of such an inquiry would be reviewed at the CoP and appropriate actions taken. Clearly, this is all very vague.

In practice, timeframes have been attached to the non-compliance procedure. It begins as outlined in Article XIII, with the Secretariat contacting the relevant Management Authority or Authorities (Wijnstekers 2018). Parties either reply within 1 month with the requested information or indicate that the request has been received and reply with an indication of when they will be able to respond in full (Wijnstekers 2018). If after 6 months, the Secretariat has not received the information that they requested, the Parties are asked to justify why this is the case (Wijnstekers 2018). The response of the Secretariat is not to go straight towards considering trade sanctions; there is always an attempt to work with parties. This usually entails an effort to address the non-compliance issue through advice and technical assistance (Wijnstekers 2018). The Secretariat also enlists help from the Standing Committee when non-compliance issues are more difficult; this may entail a more direct approach in partnership with the party to find solutions to the non-compliance (Wijnstekers 2018). All of these actions are recorded and shared with parties through official Notifications and reports for meetings of the Standing Committee and the CoP.

As mentioned, the non-compliance process has evolved outside of the text of the Convention. In particular, Resolution Conf. 14.3 (Rev. CoP18) (which means it was originally introduced in 2007 and revised in 2019), contains guidance on the compliance procedure (CITES 2019a). In the Resolution, which is still in effect, parties are informed that the Standing Committee is the main body handling general and specific matters of compliance including: "a) monitoring and assessing overall compliance with obligations under the Convention; b) advising and assisting Parties in complying with obligations under the Convention; c) verifying information; and d) taking compliance measures as described below" (CITES 2019a: 2).

The other committees of CITES—the Animals and Plants Committees—can advise and assist in this regard. In particular, if there is non-compliance related to a specific species, these committees will be tasked with a Review of Significant Trade (CITES 2019a), which, Reeve (2002) notes, is complicated. It is essentially a very detailed examination of the entire scope of trade of one species in order to determine whether trade is threatening the species' survival and why this is the case (see Chapter 5 for more detail).

The Secretariat's role in compliance is as follows:

> a) assists and supports the Animals and Plants Committees, the Standing Committee and the Conference of the Parties in carrying out their functions concerning compliance matters as described in this Guide and, where applicable, according to the procedures set out in relevant Resolutions and Decisions;
> b) receives, assesses and communicates to the Parties information on compliance mat- ters;
> c) advises and assists Parties in complying with obligations under the Convention; d) makes recommendations for achieving compliance; and
> e) monitors the implementation of compliance-related decisions.
>
> (CITES 2019a: 2)

As mentioned earlier, parties can bring to the attention of the Secretariat compliance concerns. The main means of monitoring compliance, however, are the reports that I have mentioned—annual and biennial reports—the Review of Significant Trade and the NLP (CITES 2019a).

All of this detail is to iterate that CITES is a complex treaty, which has only become more potentially unwieldy with 183 parties. The fairly low levels of implementation and compliance may also indicate that measures to improve them are ineffective. For instance, even with the NLP, implementation and compliance remain inadequate, and there is recognition that the categorizations of the NLP may not be enough, since Decision 17.60 advised for:

> Parties whose legislation is in Category 1 under the National Legislation Project are encouraged to review their national CITES-implementing

> legislation for areas where it may not fully satisfy the requirements of the Convention, in particular with regard to the possession of illegally traded specimens of CITES-listed species, and to adopt any necessary amendments. Those Parties are also encouraged to provide technical or financial assistance to one or more Parties whose legislation is in Category 2 or 3 under the National Legislation Project, either directly or through the Secretariat.
>
> (CITES 2016a: 18)

Furthermore, lack of compliance by parties "was sometimes so great that the Standing Committee delayed its determination of the failure to report for three consecutive years to its next meeting, giving Parties another year to catch up with their reporting" (Wijnsteker 2018: 350). This indicates the inability of or the lack of will by the Standing Committee to take a harder line to achieve compliance. It may also have to do with the fact that when trying to ensure compliance, the Standing Committee takes into account the circumstances of the party and there is recognition that compliance is difficult, particularly for developing countries (Wijnstekers 2018). CITES is continually trying to address compliance, as is evident in Decisions 18.62 through 18.67 from the last CoP in 2019, which were all aimed at assessing compliance and the effectiveness of implementation, and from the fact that Resolutions 11.3 and 11.17, both about compliance from 2000, are still in effect. I turn now to the Secretariat and the committees, which are spearheading these efforts.

The CITES Secretariat and committees

The CITES Secretariat is a permanent management body located in Geneva, Switzerland. They are led by a Secretary General, and all staff of the Secretariat are provided by the Executive Director of the United Nations Environment Programme (see Article XII). The Secretariat has annual management duties of drafting reports, reviewing parties' reports, and the triennial organizing of the CoP, in addition to other oversight tasks. In terms of implementation, the Secretariat has the function: "to make recommendations for the implementation of the aims and provisions of the present Convention, including the exchange of information of a scientific or technical nature" (CITES 1973: no page). As mentioned, compliance issues can be brought to the attention of the Secretariat, but they are usually dealt within the Standing Committee. Reeve (2002) has criticized the Secretariat as being too powerful (much more so than the equivalent bodies in other multinational environmental agreements), which runs the risk of (further) abuse. There is no system of review of the Secretariat's decisions or initiatives and no requirements for transparency.

The Standing Committee of CITES is where the business of the Convention takes place, including scrutiny on compliance, which may involve sanctions (CITES 1992). In other multinational environmental agreements, there are dedicated independent committees for compliance (Reeve 2002). "By allocating considerable review and recommendatory powers to the Secretariat, a non-elected body, and delegating the CoP's decision-making role on case-by-case non-compliance to the Standing Committee, CITES departs from accepted institutional requirements" (Reeve 2002: 320). Reeve (2002) acknowledges that this approach may be a strength, as it enables the use of trade sanctions. However, both the Standing Committee and the CoP, when she was writing, were reluctant to target parties for non-compliance regarding reporting and national legislation. Considering the consistent levels of partial or poor implementation and non-compliance, the same appears to be true now. Reeve (2002) also notes that the Standing Committee is a political body, which leads to several problems. First, the representation on that committee (each of the seven CITES regions are represented—Africa, Asia, Central and South America, Europe, North America, and Oceania) could act in their national interest rather than in the interests of the region. Second, the Standing Committee lacks expertise on compliance issues, so it is ill-equipped to undertake this task (Reeve 2002). She recommends an independent Compliance Committee and having the Standing Committee be responsible for oversight of the Secretariat.

Currently, there are two additional committees (see Wijnstekers 2018 for a history of CITES committees)—the Animals and Plants Committees. As mentioned, they can help with implementation and compliance in regard to species-specific issues. The scientific data they provide, which are then linked to the trade volumes and patterns, are critical to ensuring the Convention meets its goal of having trade be sustainable. With over 38,000 species listed, and the number always increasing, the task of overseeing that each species is sustainably traded is a difficult one.

CITES funding

In the original Convention text, there was no mention of how CITES would be funded (Wijnstekers 2018). During the first meeting of the Conference of Parties in June 1979, a small amendment was made to the Convention text in Article XI—CoP—paragraph 3 (CITES 1979; Wijnstekers 2018). It now reads:

> 3. At meetings, whether regular or extraordinary, the Parties shall review the implementation of the present Convention and may:
>
> (a) make such provision as may be necessary to enable the Secretariat to carry out its duties, and adopt financial provisions.
>
> (CITES 1979: no page)

The words added were "and adopt financial provisions" (CITES 1979; Wijnstekers 2018). Such a seemingly small change, though, took years to be accepted and still has what could be considered limited support from the parties. As per the regulations of the Convention, amendments have to be passed by two-thirds of the total number of parties (CITES 1979). This did not happen until 1987, when 34 of 50 parties had ratified this particular amendment (CITES 1979; Wijnstekers 2018). Currently, only 149 of the 183 parties have ratified the financial provisions amendment (CITES 1979—the website of the original text has the latest number of parties).

As mentioned, there seems to be limited support for the need for financial provisions, with just over 80 per cent of parties having adopted the amendment. This raises the question then as to how CITES is financed. Wijnsteker's (2018) *Evolution of CITES* provides a detailed history of where funds have been gathered from. To summarize, since CITES is affiliated with the UN Environment Programme, some funding originally came from there, but this gradually reduced during the first years of CITES with the expectation that parties would take on the costs of the Secretariat as well as the CoP. In the period before the Bonn amendment and the transfer of financial responsibility for CITES to the parties, the UN Environment Programme established a Trust Fund that parties and other stakeholders could contribute to (Wijnstekers 2018). The CITES Trust Fund still exists today and is where each party's contributions (and other contributions) are held. Every 3 years, at the CoP, there is a resolution that sets out the expected work programme and expenditures for the time up until the next Conference.

The latest is Resolution Conf. 18.1—Financing and the costed programme of work for the Secretariat for the triennium 2020–2022 (CITES 2019b). Resolution Conf. 18.1 refers to documents shared with the parties outlining the expenditures of the Secretariat for the previous year and the proposed costs of the work programme. Furthermore, mention is made of the increased number of parties, the increased number of species listed in the appendices, and the need to improve implementation of the Convention, all of which place greater pressure on the expenditure and the Secretariat (CITES 2019b). This is all framed within the recognition that CITES has global importance and is under more scrutiny in terms of the Sustainable Development Goals, the Aichi Biodiversity Targets, and the UN General Assembly—in particular the resolutions on tackling wildlife trafficking and creating World Wildlife Day. For the triennium at the time of writing,

> DECIDES that the implementation of the costed programme of work for the triennium 2020–2022, Annex 1, shall be covered by the Trust Fund budget (CTL), Annexes 2 and 3, in the amounts of USD 5,907,043 for 2020, USD 6,080,945 for 2021 and USD 6,697,571 for 2022 and by the Support to CITES Activities Trust Fund (QTL), Annex 3, and ADOPTS the scale of contributions for 2020–2022 as contained in Annex 5.
>
> (CITES 2019b: 3)

Parties are expected to pay their contributions in a timely manner, preferably at the beginning of the year. As detailed in Annex 5 of Resolution Conf. 18.1, each party's contribution to the CITES Trust Fund is proportionate. For example, the minimum amount a party is asked to pay for the time period between 2020 and 2022 is USD 60 per year (25 parties). In contrast, Germany's contribution is USD 1.1 million, Japan's USD 1.5 million, China's USD 2.1 million, and the US's USD 4.1 million. Overall, the 3-year budget is USD 18,685,560. There are two shortcomings to this. First, the budget includes contributions from all 183 parties, when only 149 have adopted the financial provisions amendment. Second, such financial provisions are part of a resolution, which makes them non-binding. This is evident from the language, which "requests", "welcomes", and "encourages".

Some additional funding is gained from charging observers to attend the CoP at the rate of USD 600 for the first delegate and USD 300 for each additional person (CITES 2019b). Observers are charged USD 100 for attending any committee meetings. In regard to committee meetings, the Resolution:

> DIRECTS the Secretariat to make provision for the payment, if requested, of reasonable and justifiable travel expenses of members, including attendance at the relevant committee meetings, and other expenses of the Chairs of the Standing Committee, the Animals Committee and the Plants Committee, other than members from developed countries.
>
> (CITES 2019b: 4)

Importantly then, a party's limited finances should not preclude them from participating in the Convention's various activities.

I have spent time detailing how CITES is financed because, as I will return to numerous times, financial limitations are mentioned over and over again as a challenge to both implementation of and compliance with CITES, as well as its enforcement.

A note on enforcement

It is evident from the details given earlier that there is neither an Enforcement Authority required nor is there a committee for enforcement. In practice, there is no formal way for enforcement representatives to participate in CITES (Reeves 2002). There are one or more Secretariat staff concerned with enforcement. Reeve (2002) refers to enforcement as the Achilles heel of CITES. This is for a variety of reasons. Difficulties in enforcing CITES arise from parties' national financial limitations; they do not have the funds to dedicate to enforcing permit requirements and checks or to train or equip staff (border agents, customs officers, and rangers among others). Furthermore, there is an overall insufficient amount of scientific information about the listed species. Having in-depth knowledge of the number of individuals, the number of populations, reproductive rates, trade levels, and so forth, can be challenging. In addition, for those parties which have not managed to transpose adequate legislation and regulatory mechanisms, enforcement is impossible. In terms of law enforcement, CITES and wildlife conservation remain low on the priorities of many agencies and national governments, so the list of difficulties can be a vicious cycle—no interest, no funds, no enforcement, and so forth. There is also a lack of public awareness, so individuals and organizations may be violating CITES unknowingly. "The lack of a committee dedicated to issues concerning compliance, implementation and enforcement is a significant gap in the institutional makeup" (Reeve 2002: 269).

Protectionism versus Utilization

I have said several times that the purpose of CITES is to ensure trade does not threaten the survival of the species who are being traded. This seemingly straightforward aim is fraught with philosophical and ideological tensions. Ultimately, wildlife conservation and the trade of wildlife that is entangled within conservation is an emotive issue (Reeve 2002; Challender and MacMillan 2019). There are those parties and organizations that advocate for sustainable use of wildlife, and that use can and does financially support conservation initiatives and provide livelihoods for people. The underpinning ideology seems to be an anthropocentric one, where wildlife is a natural resource that humans have the right to use and consume. The most dogmatic of the sustainable use proponents "tend to exclude ethical consideration, such as whether consumptive exploitation of certain species should be taboo" (Reeve 2002: 15). Others advocate for a more precautionary approach when species' populations decline, strict trade restrictions, and/or introduction of bans. And there are those individuals and organizations arguing for the end of wildlife trade altogether, driven by the belief that individual wildlife have the right to life (Sollund 2019; World Animal Protection 2020). As Reeve (2002: 14) puts it:

> Its [wildlife trade] control has been justified scientifically, economically, ethically, and aesthetically, with viewpoints ranging from overtly anthropocentric to overtly biocentric. Conservationists are often placed in two camps: the so-called "protectionists", who believe wildlife should be protected for its own sake, and those who promote "sustainable use", i.e., the consumptive use of wildlife at a sustainable level as a means to conserve it. In reality this is an over-simplification, and there are many shades of grey opinions.

In general, I agree with Reeve (2002: 15) that "Few wildlife policy-makers are willing to tackle head-on such issues as whether we have the right to exploit all species, or whether some should simply be left alone and, if so, where the line should be drawn".

The battle to have these competing beliefs, expressed in the Convention through its species listings, Decisions, and Resolutions, plays out at every CoP (and daily on Twitter). Listings are often made for political, economic, philosophical, and even emotional reasons, as well as scientific ones (Challender and Macmillan 2019; Vincent et al. 2013; Webb 2013). A full discussion of this tension is largely beyond the scope of this book and was largely beyond the scope of the project on which this book is based. It is, however, a core tension for CITES, so I end up returning to this topic in different ways several times. I mention wildlife protectionism versus utilization here when discussing how CITES works because, I suggest, there is a disconnect between what the Convention says and what the Convention does. The text lays out a scientifically based decision-making process for protecting species in trade from being overexploited. Yet, the process is frequently underpinned by the cultural, emotional, and political dimensions as much as the science. This is essential to understand when analysing the species who are listed and not listed and to understand the relationships between different parties to each other and to the Secretariat.

For example, at the 18th CoP in August 2019, a proposal to stop the export of wild-caught live African elephants was agreed upon. This was applauded by the African Elephant Coalition (who initiated the proposal) and other supporting parties, like the EU (Fondation Franz Weber 2019). Advocates for the elimination of the trade in live elephants cite the trauma elephants experience: from being captured and separated from their families, enduring long difficult transportation, often from Africa to China, to a life in captivity that cannot meet their biological and social needs (Fondation Franz Weber 2019). The proposal is specific to Botswana and Zimbabwe, who had been exporting wild-caught live baby elephants to zoos in Asia (Humane Society International/Global 2019), despite elephants being an Appendix I species, who should only be traded in special circumstances. Trade can happen

> only if approved by the CITES Animals Committee, in consultation with the IUCN African Elephant Specialist Group, an expert group that has publicly stated it does not believe there to be conservation benefits to wild caught elephants being sent to captive facilities. The amended text was passed by a vote of 87 in favour, 29 opposing and 25 abstaining.
>
> (Humane Society International/Global 2019: no page)

In contrast, opponents of stopping the trade in live elephants argue that decisions were made based on emotions and without considering the impact wildlife have on rural Africans (Nhongo 2019). Frustration at this ban and that the selling of stockpiled rhinoceros horn and elephant ivory would not be allowed led to the southern African nations meeting to discuss withdrawing from CITES (Nhongo 2019).

This situation demonstrates the tensions in CITES that influence how it works in practice rather than how it might be interpreted to work if relying only on the Convention text. I return to this topic in Chapter 7, when speculating about the future of CITES.

Conclusion

CITES is a complex apparatus with 183 parties and hundreds of authorities and thousands of staff working to process and monitor hundreds of thousands of permits in order for wildlife listed in CITES to be traded. Parties are required to transpose the Convention into their national legislation; in particular, they need to appoint at least one Management and one Scientific Authority, prohibit violations of the Convention, penalize those violations, and have a process of confiscating illegally traded and/or possessed wildlife. Parties are also required to send the Secretariat annual reports of all incidents of trade and biennial reports on implementation.

Despite the NLP, a Model Law, and guidance on implementation, implementation of the Convention in the parties' legislation remains inadequate with half of the parties failing to implement the Convention fully. Compliance, too, is less than ideal, which is particularly problematic since decisions regarding trade are made on the basis of the annual reports. Poor compliance is despite the fact that CITES, unlike other multinational environmental agreements, has the ability to sanction parties with trade restrictions for lack of compliance. This is rarely used though. The work of the Secretariat and the committees is confined within a tight budget—around USD 6 million per year—which is funded through non-compulsory proportionate contributions of the parties. On top of the complex bureaucracy and logistics of implementation and compliance are the equally complex cultural-economic-ethical-political dynamics that underpin CITES' decisions and processes. These need to be acknowledged and understood as they, too, play a role in implementation and compliance.

It is critical for the survival of many CITES-listed species that they are properly protected and not overexploited by trade and trafficking. This book explores in depth what can be done to improve implementation and compliance by exploring best practices and lessons learned of CITES' parties. The next chapter on methodology details how I went about gathering the data to do this exploration.

References

Challender, D. and MacMillan, D. 2019. Investigating the influence of non-state actors on amendments to the CITES appendices. *Journal of International Wildlife Law & Policy*. 22 (2): 90–114. https://doi.org/10.1080/13880292.2019.1638549.

Convention on the International Trade in Endangered Species of Wild Fauna and Flora (CITES). 2019a. Conf. 14.3* (Rev. CoP 18). CITES compliance procedures. Available at: www.cites.org/sites/default/files/document/E-Res-14-03-R18.pdf. Accessed 17 July 2020.

———. 2019b. Financing and the costed programme of work for the Secretariat for the triennium 2020–2022. Available at: https://cites.org/sites/default/files/document/E-Res-18-01.pdf. Accessed 10 December 2020.

———. 2016a. Decisions of the conference of the parties to CITES in effect after its 17th meeting. Available at: https://cites.org/sites/default/files/eng/dec/valid17/E17-Dec.pdf. Accessed 17 July 2020.

———. 2016b. National laws for implementation of the convention. Available at: https://cites.org/sites/default/files/eng/com/sc/66/Inf/E-SC66-Inf-19.pdf. Accessed 26 July 2016.

———. 2001. SC45 Doc. 21. Conservation of and trade in specific species.

———. 1992. Resolution Conf. 11.1 (Rev. CoP16), Establishment of committees. Available at: https://cites.org/eng/res/11/11-01R16.php. Accessed 26 July 2016.

———. 1979. Bonn amendment to the text of the convention. Available at: https://cites.org/eng/disc/bonn.php. Accessed 10 December 2020.

———. 1973. Convention on international trade in endangered species of wild fauna and flora. Available at: www.cites.org/eng/disc/text.php#IX. Accessed 16 July 2020.

Finnish Ministry of the Environment. 1995. *Nordic Research Project on the Effectiveness of International Environmental Agreements Draft Report*. Helsinki: Finnish Government.

Foundation Franz Weber. 2019. Press release: Historic CITES decisions prevent the export of more than 30 baby elephants in Zimbabwe. Available at: https://africanelephant-journal.com/press-release-historic-cites-decisions-prevent-the-export-of-more-than-30-baby-elephants-in-zimbabwe/. Accessed 20 July 2020.

Human Society International/Global. 2019. Momentous decision by CITES to limit trade in wild caught African elephants from Zimbabwe and Botswana for zoos to "exceptional circumstances" with independent oversight. Available at: www.hsi.org/news-media/cites-cop18-compromise-wild-caught-elephant-trade/. Accessed 20 July 2020.

Mitchell, R. 1996. Compliance theory: an overview. In Cameron, J., Werksman, J. and Roderick, P. (eds.). *Improving Compliance with International Environmental Law*. Oxon: Earthscan: 3–28.

Nhongo, K. 2019. Southern African nations threaten to leave global wildlife pact. *Bloomberg*. Available at: www.bloomberg.com/news/articles/2019-08-27/rhino-snub-draws-threat-of-africa-exit-from-global-wildlife-pact. Accessed 20 July 2020.

Raymakers, C. 2002. Biodiversity and wildlife trade: From fauna and flora in trade to sanctions. In Anton, M., Dragffy, N., Pendry, S. and Young, T. R. (eds.). *Proceedings of the International Expert Workshop on the Enforcement of Wildlife Trade Controls in the EU, 5–6 Nov 2001*. TRAFFIC/IUCN.

Reeve, R. 2002. *Policing International Trade in Endangered Species: The CITES Treaty and Compliance*. London: Earthscan.

Sollund, R. 2019. *The Crimes of Wildlife Trafficking: Issues of Justice, Legality and Morality*. London: Routledge.

Vincent, A. C. J., Sadovy de Mitcheson, Y, Fowler, S. L. and Lieberman, S. 2013. The role of CITES in the conservation of marine fishes subject to international trade. *Fish*. 14: 1–28.

Webb, G. J. W. 2013. *Wildlife Conservation: In the Belly of the Beast*. Darwin, NT, Australia: Charles Darwin University Press.

Wijnstekers, W. 2018. *The Evolution of CITES: A Reference to the Convention on International Trade in Endangered Species of Wild Fauna and Flora, 11th Edition*. International Council for Game and Wildlife Conservation. Available at: www.cic-wildlife.org/wp-content/uploads/2018/05/the-evolution-of-cites-2018-11th-editionpart-1.pdf. Accessed 27 November 2019.

World Animal Protection. 2020. End the global wildlife trade. *Forever*. Available at: www.worldanimalprotection.org/take-action/end-global-wildlife-trade-forever. Accessed 10 December 2020.

Yeater, M. 2002. Enforcement and the CITES national legislation project. In Anton, M., Dragffy, N., Pendry, S. and Young, T. R. (eds.) *Proceedings of the International Expert Workshop on the Enforcement of Wildlife Trade Controls in the EU, 5–6 Nov 2001*. TRAFFIC/IUCN.

3

METHODOLOGY

As mentioned, according to Reeves (2002), there has never been an empirical evaluation of the implementation of CITES. Furthermore, there does not seem to have been an exploration of parties' compliance with CITES either. My research project attempted to fill part of that gap, and this chapter details how I went about empirical data collection related to CITES' implementation and compliance. This project was a legal studies project grounded in the investigation of how the requirements of CITES are transposed into the national legislation of the parties (implementation) and how the parties meet the reporting requirements of the Convention (compliance). Specifically, because of the biodiversity crisis and the role that wildlife trade and trafficking play in that crisis (Chapter 1) and the state of knowledge of apparently poor implementation of and compliance with CITES (Chapter 2), I set out to achieve these three aims. First, there is a need to have more in-depth knowledge of the nature and extent of implementation of and compliance with CITES legislation in all parties to fully understand the impact of legal and illegal trade on species survival. This better understanding may also help to enlighten broader research gaps related to environmental governance in general. Second, it is important to identify lessons learned in relation to CITES legislation, implementation, and compliance in order to provide evidence to parties as to how their own practice can be improved and to inform other multilateral environmental agreements (MEAs) of approaches to implementation and compliance. And third, my aim was to widely share three case studies of best practice and lessons learned with relevant stakeholders and policy makers in order to inform changes to CITES legislation, implementation, and compliance and wider environmental governance.

To undertake such an investigation, I designed and conducted a mixed-method multistage research project. My project first consisted of content analysis of the parties' national legislation transposing CITES, and analysis of the biennial reports of the parties. Once that overall picture was obtained and after the different approaches to transposition were analysed, I consulted experts as to which of these approaches were effective and from which of these approaches the global community could learn how to improve their implementation of CITES. This was accomplished through a Delphi iterative survey. To partly assess compliance, a research assistant (Monica Pons Hernandez) compiled a chart of biennial reports that have been received by the CITES Secretariat. The combined data from the legislative content analysis, Delphi iterative survey, and review of CITES reports generated three case studies—Canada, Indonesia, and South Africa. I, then, conducted semi-structured interviews with experts about these three parties and CITES in general to gain further understanding of best practices and lessons learned. This chapter outlines these methods in greater detail later. This is followed by the justifications for designing the study in this manner. The limitations of this design are then discussed.

The legislative content analysis

The legislative content analysis was the most time-consuming stage of this 2-year project, taking approximately 1 year to complete. The first step was to identify the legislation that transposes the text of CITES into each of the 183 parties' domestic legal system. This was tackled in alphabetical order with CITES' reports, the Legal Atlas database, and the grey literature as a place to start. Sometimes, a parties' legislation will be named in a report to CITES. For instance, in the biennial reports, Question 2 asks for the title and date of any CITES-relevant legislation that has been planned, drafted, or enacted. This is not always filled in. A further question asks if the legislation has been attached to the biennial report. In cases where the "yes" box is ticked, the legislation is not always attached in the version of the biennial report available at the CITES website. The Legal Atlas database is a web-based platform designed to "use the power of artificial intelligence, maps and visual analytics to dramatically improve access to, comparison, and understanding of laws around the world" (Legal Atlas 2018: no page). The platform has 13 topics where legislation is being compiled and summaries and/or analyses given. Wildlife trade is one of these. At the time of my legislative content analysis, legislation was available for 30 parties, but the analysis did not always correspond to my specific interests regarding implementation and compliance. The grey literature that was consulted consisted of reports produced by non-governmental and intergovernmental organizations such as the World Wildlife Fund, TRAFFIC, and the UN Office on Drugs and Crime (UNODC). The reports by DLA Piper (2014 and 2015) were useful as they contained in-depth analysis of 26 parties' legislation.

The spreadsheet that I created lists the parties in alphabetical order along with their region (Africa, Asia, Central and South America and the Caribbean, Europe, North America, and Oceania) and the NLP Category (1, 2, 3, or unassessed) of their legislation as well as the title(s) of the specific legislation implementing CITES. The entire spreadsheet may be found at my website—drtwyatt.weebly.com. When the specific law (or laws) was not immediately apparent from the sources initially consulted, Google internet searches were used to identify the relevant legislation.

Searches would entail key words such as "CITES legislation", "wildlife trade legislation", and/or "wildlife legislation" and the name of the party.

Once the specific law was recorded, I then used Google Chrome and the Google search engine to search for a website that had the entire text of the legislation. These URLs, which were active when I was searching and analysing the legislation between May 2018 and May 2019, can also be found in the spreadsheet on my website. I found that CITES legislation or review of the legislation (by Legal Atlas or DLA Piper 2014, 2015) is available in English for 112 member countries. I have working knowledge of Spanish and Russian, which enabled content analysis of a further 20 parties. I used Google Chrome specifically as the web browser when searching and reading the legislation because of the automatic translate function that supported analysis of a further 47 parties' legislation. Google Chrome does not have this function for PDF documents, so in those cases I copied and pasted the text into Google translate. In four instances, where the legislation was a PDF that was protected and could not be copied, the text could not be translated (three in Arabic; one in Somali).

As mentioned, as a starting point for the legislative content analysis, I drew on the CITES NLP and the Model Law that provides guidance on the four main required components of the legislation: (1) designation of national authorities (Management and Scientific); (2) prohibition of trade in violation of the Convention; (3) penalization of illegal trade; and (4) authorization to confiscate specimens illegally traded or possessed. In regard to the Management and Scientific Authorities, I relied on the CITES website, which lists the contact information for each of the parties' Management and Scientific Authorities (if they have them). These were each recorded in a separate sheet of the spreadsheet and the email information extracted into a contact list for the second stage of the project—the Delphi iterative survey. I analysed whether the authorities were part of the same agency or whether they were separate and listed if they had an Enforcement Authority or not (and whether it was a separate entity or in the same agency as one of the other authorities). As mentioned, this is not a requirement, but is likely to affect implementation and compliance. I, then, have columns in the spreadsheet for prohibition, penalties, and confiscation. Often, I directly copied and pasted text from the parties' legislation to the relevant column.

My analysis was not confined to these four components. From reading the literature (in particular, Reeve (2002)) and my previous research on wildlife trafficking, there is evidence that other factors may contribute to the best practice regarding CITES implementation and compliance as well as limit a party's ability to fully engage with the Convention. As such, I also recorded the type of legislation transposing CITES (administrative, civil, environmental, trade, or wildlife) and if it was CITES-specific or part of broader legislation. With the help of Alison Hutchinson, we also recorded how "wildlife" was defined in each piece of CITES legislation. This included whether wildlife was broadly defined as fauna and flora and/or animals and plants, or whether wildlife was narrowly defined and excluded certain taxa of wildlife. I also recorded any information in the legislation being analysed of regulation of domestic trade and hunting, collecting, and/or harvesting. Even though CITES does not oversee domestic wildlife regulation, clearly there is a relationship between provisions within a country and the international trade of wildlife. In addition, as Reeve (2002) observed, it seems that parties with well-managed domestic wildlife also are likely to have well-managed international provisions. CITES allows parties to have stricter measures than the Convention, so I made note of this in order to see if this might affect overall implementation and compliance.

The main instrument for facilitating CITES trade is the permit system, so I recorded any information about issuing of permits and whether or not these might have differed from the required import and export permits. Connected to which permits are needed is whether or not the wildlife being traded is wild-caught or captive-bred. Appendix I species who are captive bred are treated as Appendix II listings, so would only need an export permit (rather than both an export and an import permit). This is an important distinction, and it is vital to know whether the facility that is carrying out the breeding or artificial propagation is registered. Thus, I made note of whether or not the legislation indicates there is a system in place for registering facilities that captive-breed or artificially propagate CITES-listed wildlife.

As indicated, there is not a requirement for an Enforcement Authority. It is interesting to understand then which agencies within a party are tasked with enforcing the CITES legislation in the absence of a specific agency being given the responsibility. Thus, I recorded which agencies were mentioned as enforcing CITES legislation and (if possible) what powers of authority they have. The latter speaks to the often-lesser status of environmental and wildlife law enforcement, where they might not be given full law enforcement authority to arrest suspects, investigate crimes, and/or confiscate evidence. When this is the case, it is crucial that there are both formal agreements and good relationships between agencies that have to work together to handle CITES violations. For instance, in the UK, the Border Force agents do not have full powers to arrest and investigate, so when a CITES violation is identified, either the local police constabulary has to be called in for cases appearing to be minor or the National Crime Agency must be called in cases appearing to be serious. As mentioned, I kept track of whether or not violations were penalized, since this is one of the four main components of implementation. I did so by recording what the specific penalties are in cases of CITES violations (i.e. amount of fines, prison sentence, or alternatives). These were not always immediately apparent, as penalties are often located in separate legislation.

In terms of confiscation (the last component of implementation), I took note of not only whether or not there are provisions to confiscate wildlife and wildlife products, but also whether or not the authorities can seize or confiscate other items. For instance, it may be best practice to also have the capacity to seize equipment, vehicles, and/or tools used in the violation of CITES as well as provisions for the authorities to seize or forfeit assets and proceeds of crime linked to the CITES violation. Linked to the confiscation of wildlife and wildlife products is what happens to them afterwards. For confiscated wildlife products, it is important that there is secure evidence storage. For live wildlife, there are numerous issues to contend with. In the first instance, CITES recommends returning illegally traded or possessed wildlife to their country of origin. If this is not possible, then arrangements should be made for the wildlife to go to a rescue centre or sanctuary. It is allowed for wildlife to be euthanized. I recorded what information is available in the legislation as to what happens once wildlife and wildlife products have been confiscated. I also made note of any welfare provisions within the parties' CITES legislation. This is relevant not only to confiscations of live wildlife, but also to the transport of live wildlife during trade. Welfare requirements for live wildlife are particularly relevant when considering disease transmission and the potential links to public health. The welfare of the wildlife during transport is the only welfare requirement of the Convention. Finally, in regard to the legislative content analysis, I took note of any elements of a parties' legislation that seemed to be unique or that held promise for being applicable in different socio-economic contexts.

Once this extensive spreadsheet was complete, further analysis was undertaken. The main focus of the analysis were the four components required for implementation. This analysis was fairly straightforward, as for each it is essentially a "yes" or "no" answer as to whether the required component appears to be present or not. This was then compared to the Category assigned to the parties' legislation through the NLP. The findings of this analysis are in Chapter 4. Additional analysis involved identifying patterns and trends related to relationships between the authorities, the types of penalties used, approaches to captive breeding and artificial propagation registries, and the types of welfare provisions in place, amongst other topics. Another source of information about implementation comes from the parties themselves through their biennial reports. These shed light, in particular, on the limitations and challenges to being able to fully and effectively implement CITES.

The biennial reports

On the CITES website, biennial reports are available from 2003. Actual submissions are low when considering how many there should be were each party to report as required. Monica Pons Hernandez, a research assistant, undertook the following data collection from the reports that were available. For the reports between 2003 and 2015, she pulled out answers from questions that I specifically identified and recorded them in a spreadsheet. I chose the specific questions based upon their relevance to implementation and compliance. The first question—B5[1]—is:

Next, to explore the permit process, Monica extracted the data from question D5.2 asking "To date, has your country developed written permit procedures for any of the following? Permit issuance/acceptance, registration of traders, and registration of producers where the tick box answers available were yes, no, or no information".

In order to investigate the efforts parties are making to improve implementation, one question (D6.1) about capacity building was analysed.

5	Which of the following issues are addressed by any stricter domestic measures adopted for CITES-listed species (in accordance with Article XIV of the Convention)?						Tick all applicable
		The conditions for:			The complete prohibition of:		
	Issue	Yes	No	No information	Yes	No	No information
	Trade	☒	☐	☐	☐	☐	☐
	Taking	☐	☐	☒	☐	☐	☐
	Possession	☐	☐	☒	☐	☐	☐
	Transport	☒	☐	☐	☐	☐	☐
	Other (specify)	☐	☐	☐	☐	☐	☐
	Additional comments						

FIGURE 3.1 Example of Question 5 from a CITES Biennial Report

D6 Capacity building

1	Have any of the following activities been undertaken to enhance effectiveness of CITES implementation at the national level?			Tick if applicable
	Increased budget for activities	☐	Improvement of national networks	☒
	Hiring of more staff	☐	Purchase of technical equipment for monitoring/enforcement	☐
	Development of implementation tools	☐	Computerization	☐
	– Other (specify)			☐

FIGURE 3.2 Example of Question D6 from a CITES Biennial Report

The last section (D8) of the pre-2015 biennial reports asks about Areas for Future Work. Monica recorded all the answers to the final questions, so the limitations and challenges to implementation could be analysed.

In the case of the reports between 2015 and 2018, when Monica was gathering data from the contents, I asked for the answers to the following questions to be compiled in the spreadsheet:

The revamped biennial reports (now implementation reports) are focused specifically on implementation. The questions are aligned to specific objectives and to specific indicators measuring if those objectives have been met. In the case of Questions 1.3.1 a and b, the objective (1.3) is "Implementation of the Convention at the national level is consistent with decisions adopted by the Conference of the Parties" (CITES 2018: 3). As is evident, the questions are directly relevant to implementation. Further investigation of the biennial reports focused on questions related to penalization of violations of the Convention. Question 1.7.3a asks yes, no, or no information to "Do you have law and procedures in place for investigating, prosecuting, and penalizing CITES offences as a crime? If 'Yes', please provide the title of the legislation and a summary of the penalties available" (CITES

2	Were any difficulties encountered in implementing specific Resolutions or Decisions adopted by the Conference of the Parties?	Yes No No information	☐ ☒ ☐
3	If Yes, which one(s) and what is the main difficulty?		
4	Have any constraints to implementation of the Convention arisen in your country requiring attention or assistance?	Yes No No information	☒ ☐ ☐
5	If Yes, please describe the constraint and the type of attention or assistance that is required. A computerize system for the customs offices in border crossing points throughout the country as well as the computerization of the permit issuance procedure.		
6	Have any measures, procedures or mechanisms been identified within the Convention that would benefit from review and/or simplification?	Yes No No information	☐ ☒ ☐
7	If Yes, please give a brief description.		
8	Please provide details of any additional measures taken:		

FIGURE 3.3 Example of questions from a CITES Biennial Report

1.3.1a	Has your country responded to all relevant special reporting requirements that are active during the period covered in this report, including those in the Resolutions and Decisions of the Conference of the Parties, Standing Committee recommendations, and Notifications issued by the Secretariat (see [link to location on the CITES website where the reporting requirements are listed])? Responses provided to ALL relevant reporting requirements ☒ Responses provided to SOME of the relevant reporting requirements ☐ Responses provided to NONE of the relevant reporting requirements ☐ No special reporting requirements applicable ☐
1.3.1b	Were any difficulties encountered during the period covered in this report in implementing specific Resolutions or Decisions adopted by the Conference of the Parties? Yes ☐ No ☒ If 'Yes', please provide details of which Resolution(s) or Decision(s), and, for each, what difficulties were / are being encountered?

FIGURE 3.4 Example of Questions 1.3.1a and b from a post-2015 CITES Implementation Report

2018: 9). Question 1.7.3b asks if criminal offenses of poaching and wildlife trafficking fit the United Nations Convention on Transnational Organized Crime (UNTOC) definition of serious crime, which means that the crime is considered serious enough to warrant punishment of imprisonment of over 4 years (yes, no, or no information) (CITES 2018).

Monica also collected data from the biennial reports related to confiscations, particularly of live wildlife. Question 1.7.5d is:

Objective 2.2 is "Sufficient resources are secured at the national and international levels to ensure compliance with and implementation and enforcement of the Convention" (CITES 2018: 13). I was particularly interested in Question

1.7.5d	How were any confiscated specimens disposed of?	Tick all that apply
	– Return to country of export	☐
	– Public zoos or botanical gardens	☒
	– Designated rescue centres	☒
	– Approved private facilities	☒
	– Euthanasia	☐
	– Other (please specify):	☐
	Have you encountered any challenges in disposing of confiscated specimens? **Costs are often a problem**	
	Do you have good practice that you would like to share with other Parties?	

FIGURE 3.5 Example of Question 1.7.5d from a post-2015 CITES Implementation Report

2.2.2d	What is the respective level of priority for enhancing the effectiveness of CITES implementation at the national level through the following activities?				
	Activity	High	Medium	Low	Not a Priority
	Hiring of more staff	☒	☐	☐	☐
	Development of implementation tools	☐	☐	☒	☐
	Purchase of new technical equipment for implementation, monitoring or enforcement	☐	☐	☒	☐
	e-permitting	☐	☐	☐	☒
	Other (please specify): **awareness in judiciary**	☒	☐	☐	☐

FIGURE 3.6 Example of Question 2.2.2d from a post-2015 CITES Implementation Report

2.2.2a: "Have any of the following activities been undertaken during the period covered in this report to enhance the effectiveness of CITES implementation at the national level?" The tick box answers were hiring more staff, development of implementation tools, purchase of technical equipment for implementation, monitoring or enforcement, and/or other, specifying what that is. Relatedly, Monica recorded the answers to Question 2.2.2c, about what development assistance for the CITES authorities is needed, and Question 2.2.2d:

Objective 3.4 and the associated questions go to the heart of improving CITES and the purpose of my project. Question 3.4.1a asks whether the party has any data showing a CITES species' conservation status in their country has stabilized or improved. Similarly, Question 3.4.1b asks, "Do you have specific examples of success stories or emerging problems with any CITES listed species? If, 'Yes', please provide the details" (CITES 2018: 21). Finally, Monica took note of the general feedback at the end of the biennial reports, which asks for details of constraints to implementation and examples of good practice.

Not all of the data that were collected from these reports are included in the upcoming chapters. These reports, particularly the post-2015 version, are a rich source of data and should be scrutinized in much more detail in the future for a range of aspects not only of implementation, but also of CITES enforcement. The biennial reports are important for another reason beyond their insights into implementation of CITES. Since they are a mandatory report of the Convention, they are one indication of parties' compliance with CITES. Thus, as part of the analysis of the contents of the biennial reports, Monica also recorded the number of missing and submitted reports in order to have some indication of compliance by the parties (see Chapter 5).

The combined information that was gathered from the legislative content analysis of parties' legislation and the analysis of particularly relevant aspects of their biennial reports formed the basis for creation of the Delphi iterative survey, stage 2 of this research project.

The Delphi iterative survey

The aim of the Delphi iterative survey is to provide evidence to parties as to how their own practice can be improved. As a group facilitation technique that builds consensus from experts on complex issues such as policy, a Delphi iterative survey is a strong methodology to identify best practice (De Roe et al. 2016). The survey is iterative because after the first round, the second round (and third and so forth) is a new set of questions based upon the patterns of answers observed in the previous round. Thus, consensus is built through feedback on the anonymous answers of all of the experts.

Northumbria University has a subscription to JISC online surveys, formerly Bristol Online Surveys. This password-protected website was used to create the survey. The first page was the informed consent and agreement to participate (see Appendix A for a blank Round 1 survey and Appendix B for a blank Round 2 survey; see my website for the raw data). Including the participation questions, the Round 1 survey consisted of 30, often multipart, questions. Most questions were on a Likert scale (1 = Strongly Disagree to 5 = Strongly Agree); others were ranking questions of most important element to least important, and there was the opportunity for open text comments to provide expanded thoughts and experiences on the questions. The types of questions in Round 2 were very similar and consisted of 34 questions in total. Participants in Round 2 had access to the full survey results from Round 1.

To recruit participants, I enlisted a combination of direct solicitation of the relevant stakeholders and purposive sampling of experts within my network. The direct solicitation of stakeholders links to the list of Management and Scientific authorities that I created from the CITES website that I mentioned earlier. Added to this list were CITES experts working for universities, non-governmental and intergovernmental organizations (UNODC, WWF, TRAFFIC, the Wildlife Conservation Society, Flora and Fauna International, Conservation International, and the International Fund for Animal Welfare (IFAW), among others), and wildlife law experts. Round 1 took place the end of May 2019 through August 2019. In total, I sent the link of my first-round survey to 664 people. I immediately received delivery errors from 113 email addresses of Management and Scientific Authorities taken from the CITES website. My intention was to conduct three "rounds", where, in each round, 40 experts from around the world would be asked to fill in the structured questionnaire. Unfortunately, participation was lower than anticipated. Even though 431 people went to the first page of the first round of the survey, only 32 people completed it.

My original plan was to also drum up participants in person in May 2019 in Colombo, Sri Lanka, at the CoP as well as to recruit people for the interviews. The Conference was re-scheduled after Sri Lanka had to withdraw from hosting the event after a terrorist bombing. While the Round 1 survey was available online, I decided to still attend the CoP of CITES, which had been moved to Geneva, Switzerland, in August 2019. This did spark a few more people to participate and enabled me to network with new people, whom I interviewed later.

Round 2 was open between 19 October 2019 and 31 December 2019. Similar to the Round 1 survey, 426 people visited the first page of the second round of the survey, but only 16 people completed it. For the second round, to try to increase participation, I had the survey translated into French and Spanish. When I emailed 551 CITES authorities and experts (the remaining number after removing from my list the non-working emails) the link to the second round of the survey, I indicated there were offline French and Spanish versions. A further 12 emails were unable to be delivered to the CITES authorities listed on the CITES website. Seven additional people reached out to complete the survey this way, thus bringing participants in Round 2 to 23 (only 21 answers are used as one person did not agree to participate even though they completed the survey and another participant only answered some of the questions).

Both surveys were completely anonymous and confidential. I did, however, end the surveys with demographic questions to gauge the representativeness of participation. For Round 1, 14 women and 16 men as well as two people preferring not to identify their gender participated. In terms of the type of organization that they worked for, 17 participants were government employees, eight from academia, four from civil society, and three from other sectors. Most participants were from Europe (13), North America (7), and Africa (6). Three participants were from Oceania and one each from Asia and Central and South America and the Caribbean. For Round 2, seven women and 13 men completed the survey and one person who preferred not to say their gender. Again, 17 participants were from government organizations, and two each from academia and other. In terms of regions, there were four participants each from Africa, Asia, North America, and Central and South America and the Caribbean. There were five participants from Europe and none from Oceania. I will discuss the limitations to this approach in a later section. Now, I elaborate on the third stage of the research, which was the identification of three case studies and interviews to expand upon these.

Semi-structured interviews and case studies

Three countries that best embody the most important lessons were identified and were the subject of further research—Canada, Indonesia, and South Africa. This was done on the basis of the responses of participants in the Delphi iterative survey rounds. Purposive sampling was used to select interview participants for each case study, who were interviewed via telephone or Skype. The interviews uncovered further insight into the specific challenges and approaches in implementing and complying with CITES. I conducted 20 interviews that lasted between 30 and 65 minutes. For some of the interviews, I took notes, and for others I used my mobile phone to make an audio recording. The digital audio recordings were uploaded to the secure website of NJC Secretarial, who completed the transcriptions. To ensure anonymity and confidentiality, interview transcripts are referred to in text by codes—A, academic; CS, civil society; and G, government—followed by a number (i.e. A1, CS2, G3, and so forth).

Whereas the CITES NLP has presumably contributed to the increase in parties that have legislation classed as Category 1 (meets the requirements for implementation), this project went a step beyond by creating specific examples through case studies of what effective legislation, implementation, and compliance might look like. To maximize accessibility of the case studies, I wrote a two-page summary of each case study. These were each translated from English into Arabic, Chinese, French, and Spanish. The summaries were shared with all Delphi iterative survey participants, interviewees, and all CITES Management and Scientific Authorities and are posted on my website.

Interviewees also discussed improvement to CITES in general. These were included in the recommendations that formed part of the Executive Summary of the whole project. This four-page document was also translated from English into Arabic, Chinese, French, and Spanish and is available on my website.

Limitations and challenges

As with any chosen research design and method, there are limitations to the approach that I adopted. In terms of the legislative content analysis, the language that the legislation is written in is an obvious limitation. Google translate addresses this to some degree, but there are, undoubtedly, translations and specific vocabularies that could be more accurate. The challenge of the legislative content analysis was its scale. Identifying and then finding versions of each piece of legislation for 183 parties is a large task in itself, which is made even more complex when considering how parties approach managing the trade in different species of wildlife. I am certain that there is forestry and fishing legislation, applicable in the CITES context, which I have not found. This is because plants, timber in particular, and fish (see Wyatt et al. Under review) are treated differently from terrestrial animals and sometimes not even legally defined as "wildlife", as I discuss further in the next chapter. I think there is even further legislative identification and analysis needed to fully understand the implementation and compliance with CITES for plant and marine species. Legislation is also a dynamic area. I presume that some of the legislation that I analysed and that I have listed in my spreadsheet has already been updated or replaced. I picture that spreadsheet as a living document, and I would welcome people to contact me with corrections and additions (it is available on my website).

The biennial (implementation) reports, as I said, are a rich source of information that should be the focus of further content analysis. Their value though is limited by two factors. First, as I will detail in Chapter 5, a small percentage of the parties actually complete them. So, the information that can be gleaned is already limited to the parties who are able to and want to share their stories of implementation of CITES. Second, those who do complete the reports do not always ensure that their answers and/or forms are complete and comprehensive. Quality of reports is an element of compliance that certainly needs to be addressed.

I admit that the main limitation of the study is the low participant engagement with the Delphi iterative survey. The 5 per cent response rate was disappointing. I can only speculate as to why this was the case. In the first instance, I think it would have helped to have the first round also available in multiple languages and for those to be online. This certainly prompted several additional participants in Round 2 and were it to be online, this may have had an even greater uptake. A further limitation is that internet may be limited in certain geographies, so not everyone who I reached out to may have had the capacity to respond.

Delphi iterative surveys are, I think, more difficult than traditional surveys to get people to take part in because they are quite lengthy and are likely to require multiple engagements due to the rounds. Since my Round 1 participation was less than hoped for, I decided to open Round 2 to previous and new participants. Only ten people responded to both surveys. This, too, is a limitation of my findings, as it misses out on some of the continuity of feedback and iteration, which is a strength of this survey method. So the length of the individual surveys and the multiple times to take part are a limitation. CITES specifically, and wildlife in general, are such low priorities, I suspect that the low response rate might also be linked to the fact that staff of the authorities have other more pressing concerns. My poor participation rates seem to reflect the inadequate implementation of and compliance with CITES.

Apart from low engagement, the biggest challenge of the Delphi iterative survey may be either uninformed responses or misleading answers. I will discuss this in greater detail in Chapter 6, when drawing out the lessons learned and best practices from the case studies, but for now I want to acknowledge that the choice of case study countries surprised several interviewees, as they believed Indonesia in particular and, to a lesser degree, South Africa to have questionable implementation of and compliance with CITES and were not examples of best practice. Both countries were named in the survey as having best practice, and in an effort to have a diverse set of case studies (rather than simply taking the three listed the most—the US, the UK and the EU), I opted for a geographical breadth and to learn more about the approaches of parties that are not covered in the literature. This led to a further limitation, one that I am consciously aware of from my other research, which is that there is no coverage of Latin America.

A sidenote from distributing the Delphi iterative survey—as I mention earlier, many of the CITES authorities' contacts were out of date and even more became out of date in the 6 months between the rounds of the Delphi iterative survey. This indicates two things. One, that the CITES website is either not routinely updated, or that parties do not keep the CITES staff up-to-date. And two, that there is a noticeable amount of turnover in CITES staff at the party level. The first implies lack of resources in the Secretariat or poor communication by the parties, both of which could be problematic. The second indicates potential trouble for implementation, compliance, and enforcement if there is no continuity of personnel working on such a complex convention.

The final note about challenges is that the final 2 months of this study fell within the early fears and then responses to the coronavirus pandemic in Europe. This meant that I was able to attend a UK House of Lords event about efforts to have wildlife trafficking as a fourth protocol to UNTOC on 3 March 2020, but my own dissemination event and plans to share my findings in person with CITES staff were not possible. As an alternative means of sharing, I recorded presentations with voice overs of the main findings, the methodology, the three case studies, and recommendations as short videos that are, again, posted to my website.

Justification and benefits

Since this study had multiple methods, this ensured that the limitations and challenges could be overcome and that I could still produce a robust data set to contribute to the improvement of CITES, in particular, and environmental governance in general. The legislative content analysis provides in-depth insight into not only implementation of CITES from an independent perspective (not CITES itself or a parties' authority), but also an evaluation of the NLP as well as an additional three core elements of implementation that I present here. This may well inform changes to CITES processes or spark interest in new initiatives to improve implementation.

The Delphi iterative survey, while a fairly small sample, is rich in information that adds depth to the current understanding of CITES implementation and compliance. The combination of set questions and open text comments provides a range of data to make a compelling case for how certain parts of CITES could be improved. The case studies, though somewhat controversial, may not always present the best practice, but there are still lessons to be learned from these experiences both for other parties and for the CITES Secretariat in regard to its relationships with parties. The open text answers to some of the Delphi iterative survey questions are also likely to lead to new avenues of inquiry and thus provide some new directions for research on wildlife trade and trafficking more broadly. The interviews and the case studies add to the robust data set by adding an additional layer of understanding of what implementation and compliance look like in practice. These rich narratives and concrete examples are lacking from the legislative content analysis and analysis of the biennial reports. Together, the mixed methods are intended to help parties further their efforts to make CITES better. By sharing summaries of the findings directly with CITES stakeholders and by making them publicly available through my website, my intention is to provide an open resource for anyone interested in CITES.

Conclusion

I undertook this project as I believe(d) that we need to understand this complex regulation and governance of international wildlife trade in more detail. The detail is important as it is likely that individual parties have best practices that should be shared in terms of ways of implementing and complying with CITES legislation. The research design, with mixed-methods and an extensive legislative content analysis, was also intended to uncover lessons learned from parties' experiences. The difference between best practice and lessons learned is that best practice has positive outcomes, whereas lessons learned are approaches that may not have positive results. However, lessons learned are a wealth of information for avoiding mistakes and misdirections. The overarching goal has been and continues to be to strengthen CITES through better implementation and compliance as well as to improve efforts to limit rule-breaking and criminality and therefore protect wildlife. The multifaceted findings that have emerged as to how implementation (Chapter 4) and compliance (Chapter 5) can be amended and added to in order to improve CITES and better protect wildlife in trade are the subjects that I turn to next.

Note

1 The screenshots of the report pictured are for Albania simply because it is the first report available on the CITES website.

References

Convention on the International Trade in Endangered Species of Wild Fauna and Flora (CITES). 2018. Implementation report 2015–2017 Austria. Available at: https://cites.org/sites/default/files/reports/15-17Austria.pdf. Accessed 23 July 2020.

———. 2005. Albania biennial report—notification no. 2005/035. Available at: https://cites.org/sites/default/files/reports/03-04Albania.pdf. Accessed 23 July 2020.

De Roe, R., Melnychuk, N., Murray, D. and Plummer, R. 2016. Advancing the state of policy Delphi practice: A systematic review evaluating methodological evolution, innovation, and opportunities. *Technological Forecasts and Social Change*. 104, Mar: 78–88.

DLA Piper. 2015. *Empty Threat 2015: Does the Law Combat Illegal Wildlife Trade? A Review of Legislative and Judicial Approaches in 15 Jurisdictions*. Washington, DC: DLA Piper and WWF.

———. 2014. *Empty Threat: Does the Law Combat Illegal Wildlife Trade? An Eleven-Country Review of Legislative and Judicial Approaches*. Washington, DC: DLA Piper and WWF.

Legal Atlas. 2018. Home. Available at: www.legal-atlas.com. Accessed 23 July 2020.

Reeve, R. 2002. *Policing International Trade in Endangered Species: The CITES Treaty and Compliance*. London: Earthscan.

Wyatt, T. Friedman, K. and Hutchinson, A. Under review. Are fish wild? *Animal Law Review*.

4

CITES IMPLEMENTATION

The big picture

This chapter—"CITES Implementation: the big picture"—is the first findings chapter. As noted by Reeve (2002) nearly 20 years ago, reports to the CITES Secretariat about implementation of the Convention by the parties indicated that the process was going slowly. Unfortunately, the big picture of implementation does not appear to have changed all that much. This chapter details the findings of the legislative content analysis of the 183 parties' wildlife trade and relevant legislation combined with the relevant responses to the Delphi iterative survey and any pertinent comments from the semi-structured interviews. Whereas CITES is touted as one of the most successful conventions, according to CITES' own measure, over half of the parties still have implementation gaps in their legislation. I begin, though, by sharing my analysis of the type of legislation that parties use to implement CITES. The chapter breaks down the nature and extent of the parties' failure to implement the four main components: designate one Scientific and one Management authority, prohibit trade in violation of the convention, penalize trade violations, and provide for confiscation of illegally traded and possessed wildlife. I also share what experts and CITES reports identify as the main limitations and challenges to implementing CITES. The CITES convention, as mentioned, is 25 different articles, so there is more to implementation than the four main components assessed by the NLP. In this chapter, I look at three further components of implementation: captive breeding registries and monitoring, permit systems, and welfare of wildlife. I conclude with some thoughts on improvements that could be made in regard to implementation.

Type of legislation

The text of the CITES convention does not stipulate how the Convention should be or needs to be transposed into a party's domestic legislation. As with the four components of the NLP that I talk about next, there is a tremendous amount of discretion given to the parties in regard to implementation. Furthermore, CITES does not require that violations of the Convention be criminalized, so I was curious to see if administrative, civil, criminal, environmental, and/or wildlife law formed the approaches of the various parties. With that in mind, I recorded what type of legislation was used to implement the Convention. As CITES is a trade treaty, it is possible that wildlife trade would be legislated under the provisions of all the various goods and services that are imported and exported by the parties. This, however, was rarely the case. I suggest that only two of the parties (Burundi and Maldives) have, at least, some of their provisions in export or import pieces of legislation.

The most common approach to implementation was within environmental law—80 parties appear to blend CITES into environmental codes. This means that CITES is woven in with other environmental legislation, such as national parks, biodiversity protection, more generally, and even pollution and waste. The next most common means of implementation of CITES was CITES-specific legislation. Several of the 48 parties that took this approach simply transpose the Convention text directly into the party's own piece of national legislation. A further 21 parties have incorporated CITES into wildlife legislation. Canada, for instance, takes this approach by having interprovincially protected species and the internationally protected species combined into a single piece of wildlife law—Wild Animal and Plant Protection and Regulation of International and Interprovincial Trade Act (WAPPRIITA) (Wyatt 2020). The remaining parties utilize a variety of other approaches ranging from environmental crime (Venezuela) to hunting (Albania, Djibouti, and Montenegro). My categorizations are not distinct, and there is certainly overlap and uncertainty. The main point is to illustrate the range

TABLE 4.0 Type of legislation

Animal	3
Animal and Plant separate	1
Blank	11
CITES-specific	48
Conservation	6
Conservation and Parks	1
Environmental	80
Environmental and Agriculture	1
Environmental Crime	2
Environmental Security	1
Forestry and Hunting	1
Forestry and Wildlife	1
Hunting	3
Trade	2
Wildlife	21
Wildlife Trade	1

of approaches and amount of discretion given to the parties as well as to set the stage for not only the upcoming discussions on the four components of the NLP in this chapter, but also the later discussions of the apparent bias in listings of CITES species (Chapter 6) and whether or not there should be a wildlife crime convention (Chapter 7).

Authorities

CITES requires there to be one Scientific and one Management Authority. As outlined in Chapter 2, CITES does not dictate how the authorities are related to each other, how the authorities should be structured, or which agencies the authorities should be. As such, there is a range of combinations of structures and agencies that make up CITES authorities in the different parties. According to Bowman et al. (2010), having two authorities was, to some degree, designed to ensure that there would be some compliance. Furthermore, the approach to the authorities was "to set up at least two bodies to enforce the Convention, to communicate regularly with other parties, to communicate regularly with the Secretariat and to meet regularly to review implementation of the Convention [and]—are all critical factors" (Bowman et al. 2010: 533). So not only is having the authorities part of implementation, but the authorities' purpose is also to further implementation.

Four countries do not have a Scientific Authority (SA) (see Table 4.1), which means that they have not properly and fully implemented CITES. It also means without a Scientific Authority that the needed scientific oversight to check if wildlife are being traded in a sustainable way that does not threaten their survival is not happening. Three of these four parties' legislation is categorized as Category 3 (not implemented). One party has not been assessed yet. In addition, Namibia technically does not have a Scientific Authority. The Namibian legislation sets out a technical committee to act as the Scientific Authority with membership from ministers, veterinarians, and other experts (Government Gazette of the Republic of Namibia 2008). Since CITES does not establish a required structure, this appears to meet the criteria for implementation since Namibia's legislation is Category 1.

Of interest, although not required in CITES, many parties have opted to have an Enforcement Authority (EA), which is an agency or agencies specifically tasked

TABLE 4.1 Breakdown of CITES Authorities[1]

Separate EA, MA, SA	42	Same EA and MA	14
Same EA, MA, SA	17	No SA	4
No EA	85	Overlap of EA, MA, SA	9
Separate MA and SA	58	Other	4
Same MA and SA	29		

1 Enforcement Authority (EA), Management Authority (MA) and Scientific Authority (SA). The total number of combinations is not 183 parties since I am showing a range of different relationships between authorities, including those who have an Enforcement Authority.

with making sure the Convention is complied with by individuals and organizations importing and exporting CITES-listed species. This might be a Customs or Border agency (i.e. the UK) or a wildlife law enforcement agency like in Canada and the United States. Eighty-five parties do not have an Enforcement Authority. This raises the issue of the level of attention given to discovering violations as well as to any resulting arrests, prosecutions, and so forth when violations do take place. In general, participants of the Delphi iterative survey agreed that CITES members should be required to have an Enforcement Authority.

Most parties have separate Management Authorities (MAs) and Scientific Authorities. And many of those with Enforcement Authorities have these as a separate authority as well. There are parties who have structured their CITES implementation to have all (two or three) authorities to be the same agency. I explored the relationship and structure aspect of authorities because I wanted to see whether there was concern that by locating the authorities within the same agency this created conflicts of interest. For instance, could it not be seen as or prove to be problematic that the same agency which was deciding how much trade was sustainable was also issuing the permits for that trade (and charging fees for those permits). My Delphi iterative survey participants, for the most part, mostly agreed that the authorities should be separate. However, they stressed that the most important aspect is that each of the two (or three) authorities are independent without pressure when making decisions and that they have clearly defined roles.

For the authorities, this quote from a Round 2 survey participant sums up the consensus:

> The keys to success for enforcement are: independent decision making for risk management, good communication with the MA (regardless of embedded or outside agency enforcing), and a values-based understanding of trade and application of law.

Prohibition

From the legislative content analysis (see Table 4.2), it first appears that a majority of parties do not have legislation that prohibits violations of the Convention as required. Fifty-seven pieces of legislation explicitly prohibit violations of CITES for all CITES-listed species. A further 45 parties' legislation seems to have partial prohibitions that do not fully include the range of species listed in the CITES appendices. For instance, Argentina's Wild Fauna Conservation Law clearly outlines that the hunting, harassment, harvest, possession, transit, use, manufacturing, and trade of wild fauna is prohibited, but this excludes fisheries. This raises (again) the importance of definitions of wildlife not only in regard to inclusion of fish, but also in terms of inclusion of plants (which I return to in Chapter 6). In this example, it is possible that both fish and plants are regulated in other legislation. This would mean that the combined legislation meets CITES' requirements, but

TABLE 4.2 Countries which prohibit violations of the Convention

Prohibition seems in line with CITES	57
Some prohibition is apparent	45
No explicit prohibition found	77
Prohibitions not in line with CITES	1
Not assessed	3
TOTAL	183

adds complexities that are difficult to implement, to comply with, and to enforce as well as difficult to assess. Other parties, which appear to have only partially implemented the prohibition requirement, also only seem to include some wildlife in their legislation.

Other trends of how prohibition seems to be incomplete is only including native species in the legislation. This means that CITES species from other countries are not protected. Malawi's laws may be an example of this. Malawi's National Parks and Wildlife Act Part I, Section 2 states: " 'wildlife' means any wild plant or animal of a species native to Malawi and includes animals which migrate through Malawi, and biotic communities composed of those species" (Waterland et al. 2015: no page). In this case, non-native species are written out of the legislation. In other instances, this exclusion may not be written down, but evident in the enforcement of CITES.

> Some countries may prohibit import in violation of CITES, without meaningful penalties or judicial action. Others only enforce CITES for native species. Should evaluate whether the legislation covers ALL CITES-listed species (not only native species; some countries don't cover marine species).
>
> (Delphi iterative survey Round 1 participant—emphasis in original)

One interviewee pointed to India as being an example of where the legislation meets the implementation requirements, but then the enforcement of the Convention only applied to native species (CS1). Another interviewee indicated something similar for Thailand (CS7). The native–non-native aspect of trade was further explored in Round 2 of the Delphi iterative survey since it had been raised in the first round. Participants disagreed that this was a problem. May be this is because only a few parties approach implementation and enforcement this way, so not all survey participants would have knowledge of some parties not protecting non-native species in the same manner as native species.

Furthermore, some legislation only addresses part of the wildlife supply chain. For instance, prohibiting the hunting or collection of CITES species, but not explicitly prohibiting import and export without permits. For instance, the legislation in Ghana—Act 43 and the Wildlife Conservation Regulations 1971, 1983—seems to prohibit only hunting, capturing, and destroying of species listed in the schedules of the legislation. In addition, CITES is clear that both trade and possession in violation of the Convention need to be prohibited. Yet some parties' legislation has prohibited one, but not the other. Congo, for instance, has legislation that prohibits poaching and trafficking. There are very clear penalties for these offences, but nothing specifically about possession of wildlife in violation of CITES (DLA Piper 2015).

An interesting aspect of implementation of CITES legislation by several parties is the focus of the text on commercial use being prohibited. In the case of Luxembourg, for example, the prohibition clearly exempts wildlife and wildlife products as personal effects. The Gambia's legislation focuses on commercial businesses not selling meat or trophies, and non-commercial exploitation is allowed in protected areas by residents. I should be clear that this is within the scope of CITES. In Article VII (Exemptions and Other Special Provisions Relating to Trade) of the Convention, the text clearly states, "The provisions of Articles III, IV and V shall not apply to specimens that are personal or household effects" (CITES 1973: no page). Article III is "Regulation of Trade in Specimens of Species Included in Appendix I", whereas Article IV pertains to Appendix II trade and Article V pertains to Appendix III trade. This means that the restrictions and prohibitions placed on CITES-listed species are not relevant to personal or household effects. However, Article VII also states:

> This exemption shall not apply where:
>
> (a) in the case of specimens of a species included in Appendix I, they were acquired by the owner outside his State of usual residence, and are being imported into that State; or
> (b) in the case of specimens of species included in Appendix II:
>
> (i) they were acquired by the owner outside his State of usual residence and in a State where removal from the wild occurred;
> (ii) they are being imported into the owner's State of usual residence; and
> (iii) the State where removal from the wild occurred requires the prior grant of export permits before any export of such specimens; unless a Management Authority is satisfied that the specimens were acquired before the provisions of the present Convention applied to such specimens
>
> (CITES 1973: no page)

Thus, the personal or household effects exemption seems to be for wildlife already within a person's possession and does not appear to be intended for acquiring CITES-listed wildlife while outside of a person's home country. The legislation, and again the enforcement, of that exemption, though, needs further examination to understand how this is practically being implemented and if this provision is in reality a loophole for people to buy and/or consume CITES-listed wildlife.

In 77 pieces of parties' legislation, I found no evidence of explicit prohibition. There is, in this case, an added layer of complexity. Prohibition can be inferred from penalties. For these 77 parties' legislation, 38 of them list penalties within

TABLE 4.3 Ways to improve whether or not countries prohibit trade in violation of CITES

Improvement or Alternative	*Strongly Agree*	*Agree*	*Neither*	*Disagree*	*Strongly Disagree*
Introduce trade suspensions against those countries which have not prohibited trade in violation of the convention	11	13	3	5	0
Introduce reduced quotas against those countries which have not prohibited trade in violation of the convention	5	11	8	7	1
Create a task force or working group to assist countries with this component	14	16	1	0	0
More visibility on CITES website and in the media about which countries do and do not prohibit trade in violation of the convention	11	13	2	6	0
Establish a system of country mentorships to have countries assist each other	10	20	2	0	0

their legislation. This paints a much more complimentary picture of implementation and would mean 75 per cent of the parties have some form of prohibition against violations (57 explicitly prohibiting, 45 with some prohibition, and 38 implying prohibition by having penalties). Further difficulty in gaining a truly accurate picture of the nature and extent of prohibition is that penalties are not always contained within the environmental, wildlife, and/or CITES legislation and may be located within another set of legislation, such as a penal code. Thus, my assessment is likely incomplete. This could be interpreted positively though and may mean that prohibition is much more widely implemented than it first appears.

In regard to prohibiting trade, for this component, participants agreed with suggestions to improve parties' implementation of CITES, which were supportive rather than punitive. I asked Round 1 survey participants:

So while there was support for trade suspensions (11 Strongly Agree; 13 Agree), reducing quotas (5 Strongly Agree; 11 Agree), and publishing on the CITES website parties which do and do not prohibit trade (11 Strongly Agree; 13 Agree), there was more support for a task force (14 Strongly Agree; 16 Agree) and a mentorship system (10 Strongly Agree; 20 Agree). Overall though, prohibition is approached very differently across parties, and there are clear gaps in terms of all species being protected.

Penalization

As mentioned, penalties are a difficult component to analyse since it is possible that the penalties for violations of the main law implementing CITES sit within another

TABLE 4.4 Penalties

Fines	16
Fine/Prison	99
Prison	3
Unclear	13
Blank	52
TOTAL	183

piece of administrative, civil or criminal legislation, or within a penal code that outlines sanctions. Therefore, while a majority of parties appear to have provisions to penalize violations of CITES, there were 65 countries for which penalties could not be determined (see Table 4.4). More than half of parties have both fines and/or prison as a means to punish CITES violations. Within these 99 parties' legislation, and the further 16 parties who only have fines for punishment, there is a great deal of variation. Fines may be set within specific ranges or may be dependent upon the income of the person who has violated the law. Fines may also be dependent upon the type of violation and, in some instances, the species, who is the target of the wildlife crime. For instance, in Albania, the fine for exporting a CITES species without proper documentation is 16,000 lek (116 GBP) whereas the fine for possession of an illegal skin is between 50,000 and 80,000 lek (364–582 GBP) (UNODC 2008). In the case of Bangladesh, there is a fine of up to 50,000 taka (543 GBP) and/or 1 year of prison if the offence involves a tiger, elephant, cheetah, hoolock gibbon, sambar deer, whale, or dolphin (UNFAO 2012). One Round 1 participant noted that fines are important, as in many parties they are used to fund conservation projects and mitigate the harm caused by the violation.

Further analysis is necessary to unpack the level of the sanctions. There has been a push to have parties increase their penalties (if they are not already to the level) to a minimum of four years of prison. This corresponds to the definition of "serious" crime that is found in the text of UNTOC. The benefit of such an approach is that making wildlife crime and trafficking a serious crime under UNTOC means that the various tools for international cooperation that UNTOC prescribes become available to the parties (UNODC 2020). So, for instance, if a party needed mutual assistance to investigate a violation of CITES trade or possession, and this party punished the violation with a sanction of more than four years of incarceration, then UNTOC provisions for help would be available. It is beyond the scope of this chapter and this book to analyse whether wildlife trafficking is a transnational organized crime, though colleagues and I have discussed this in other places that not all wildlife trafficking is organized crime and claiming that is the case creates a number of problems (Arroyo-Quiroz and Wyatt 2019; Wyatt et al. 2020).

In 2016, UNODC's *World Wildlife Crime Report* explored the number of CITES parties which had penalties that fit within the UNTOC's definition of serious crime. At the time, there were 182 parties, and they explored the penalties of the

131 parties for which there were data (UNODC 2016). Thirty-one per cent had fines only, 43 per cent less than 4 years of imprisonment, and the remaining 26 per cent had penalties of over 4 years of imprisonment (UNODC 2016). Delphi iterative survey Round 1 participants were supportive of a 4-year minimum prison sentence being made a CITES requirement (Strongly Agree 12; Agree 8—see Table 4.5). It is worth noting that some scholars warn against an overly punitive approach to offences involving animals and wildlife (and crime in general). This is due to the wealth of evidence that harsh punishments do not reduce crime or offender recidivism (see Marceau 2019), as is particularly clear from the approach to drug trafficking (Felbab-Brown 2018). Further data collection is warranted to uncover whether these penalties—be they fines or prison—are actually applied or how these violations are punished in practice.

Similar to prohibitions, Delphi iterative survey Round 1 participants agreed with suggestions that to improve parties' implementation of the penalties required by CITES, supportive rather than punitive measures should be utilized (see Table 4.5). Trade suspensions and reduced quotas were the least supported suggestions (Strongly Agree 8; Agree 1; Neither 7; Disagree 6 and Strongly Agree 6; Agree 12; Neither 7; Disagree 6; Strongly Disagree 1). As seen in Table 4.5, the task force around improving penalties across CITES parties was the most supported suggestion. After that was mentorships and then more visibility on the CITES website as to which party is or is not penalizing trade.

TABLE 4.5 Ways to improve whether or not countries penalize trade in violation of CITES

Improvement or Alternative	*Strongly Agree*	*Agree*	*Neither*	*Disagree*	*Strongly Disagree*
Introduce trade suspensions against those countries which have not penalized trade in violation of the convention	8	1	7	6	0
Introduce reduced quotas against those countries which have not penalized trade in violation of the convention	6	12	7	6	1
Create a task force or working group to assist countries with this component	11	17	4	0	0
More visibility on CITES website and in the media about which countries do and do not penalize trade in violation of the convention	10	13	4	5	0
Establish a system of country mentorships to have countries assist each other	9	18	3	2	0
Should 4-year penalties be made the standard for CITES legislation?	12	8	7	4	1

A few respondents commented on the complexity surrounding penalties. One noted:

> Clearly it [penalties for violations] needs to be proportional to the penalties for other illegal acts. But at the same time some countries impose punishment that may be viewed as either extreme or too weak to other countries. Imposing proportionate penalties in countries that have highly disparate cultures, economic classes, etc., is very challenging.

Relatedly, another participant said how penalties for crimes are often dictated by national sovereignty, so the ways in which parties can respond is taking place within set parameters.

With that said, participants in the first round of the Delphi iterative survey suggested that forfeiture of assets be considered as a penalty (this is also discussed more in the next section on confiscations). I followed up in Round 2 asking whether or not participants thought forfeiture of assets and/or confiscation was an underused CITES penalty. Twelve of the 19 participants who answered this question strongly agreed or agreed. Other Round 1 participants recommended civil as well as criminal penalties. This happens in the UK, where, for example, compliance and stop notices can be issued for CITES violations.

In Round 1, one participant suggested, "Actions need to be taken by importing countries as well—they should refuse any imports from countries without strong CITES-implementing legislation than [that] has the capability to fully penalize violations". I think this suggestion raises a number of points. First, it proposes sharing the responsibility of ensuring implementation of and compliance with CITES between not only the CITES Secretariat, but also between all of the parties. The idea of a more proactive role by parties in checking that their fellow parties are fully implementing and complying with CITES could prove an influential force in improving implementation and compliance. Second, this suggestion could be interpreted as a more supportive rather than punitive improvement mechanism. Rather than taking in wildlife shipments that might be potential violations and then possibly having to invoke penalties, if importing parties worked with exporting parties to ensure all requirements were being met, or even if exporting parties knew that there were going to be stringent checks, this could also influence improvement. In Round 2, to expand upon this point, I asked whether parties could respond more strongly by not allowing countries with poor implementation of or compliance with CITES to import CITES species. Of the 20 participants who answered this question, 15 strongly agreed and agreed to this suggestion. Finally, in terms of penalties, one participant stated "Much of this problem extends well beyond CITES itself into more systematic matters of corruption, and poor governance. Unless these issues are addressed, there will be no progress on CITES enforcement".

Confiscation

A majority of countries seem to have provisions to confiscate wildlife and wildlife products and some go further to mention equipment (i.e. tools, vehicles). I was

TABLE 4.6 Confiscation of illegally traded or possessed wildlife and the equipment

Wildlife only	71	Evidence	1
Both (wildlife/equipment)	47	None	1
Equipment	8	Blank	55

TABLE 4.7 Ways to improve whether or not countries confiscate illegally traded and possessed wildlife

Improvement or Alternative	*Strongly Agree*	*Agree*	*Neither*	*Disagree*	*Strongly Disagree*
Introduce trade suspensions against those countries which do not confiscate illegally traded or possessed wildlife	9	12	6	4	1
Introduce reduced quotas against those countries which do not confiscate illegally traded or possessed wildlife	7	7	9	8	1
Create a task force or working group to assist countries with this component	14	16	2	0	0
More visibility on CITES website and in the media about which countries do and do not confiscate illegally traded or possessed wildlife	10	11	5	6	0
Establish a global or regional network of zoos, aquariums, and gardens that assist	10	13	8	1	0

unable to determine for 55 countries what confiscation measures are in place (See Table 4.6).

The same was true for confiscations in terms of the suggestions for improving implementation, as was the case for prohibition and penalization—participants supported positive rather than negative mechanisms. Thus, a task force and a network of zoos, aquariums, and gardens to assist when wildlife is confiscated were the most supported. Admittedly, as participants pointed out, the network proposed is focused on live wildlife, so is narrow in scope as a majority of CITES trade is in wildlife products.

Participants in Round 1 of the Delphi iterative survey recommended that parties should try to utilize or draft legislation that allowed for cost recovery mechanisms for housing live wildlife and storing evidence, etc., as well as having legislation for asset forfeiture. One respondent stated:

> require member states to bi-annually report on the outcome of animals and publicise this data on CITES website, require MS [Management and Scientific authorities] to develop a long term and adequate fund to both rehome and repatriate confiscated wildlife, with priority given to repatriation where

possible. Look at opportunities to pass the cost of above onto the offender or onto non-compliant organisations. Develop partnerships with transport industry to help fund and facilitate repatriation. Use the seized proceeds of crime to fund the above.

In Round 2 of the Delphi iterative survey, I expanded upon these suggestions around cost recovery and asset forfeiture. First, I asked "Countries should consider implementing mechanisms for cost recovery in cases where any type of confiscations occur". Nineteen participants answered this question—five of them chose Strongly Agree and 12 Agree. Second, I asked "Countries should consider implementing mechanisms for asset forfeiture in cases where any type of confiscations occur". Eight strongly agreed and ten agreed (1 answered neither). My reasoning behind any type of confiscation was to address the previous observation that confiscation did not only mean live wildlife, but any wildlife product.

However, the CITES Secretariat is also concerned with what happens to live confiscated wildlife as is evident from Question 1.7.5d in the biennial (implementation) reports. This multipart question first asks, "How were any confiscated specimens disposed of? Tick all that apply" (CITES 2018: 12). The choices are: return to country of export, public zoos or botanical gardens, designated rescue centres, approved private facilities, euthanasia, and other (please specify). Question 1.7.5d also asks, "Have you encountered any challenges in disposing of confiscated specimens"? and "Do you have good practice that you would like to share with other Parties"? (CITES 2018: 12). As I discuss in greater depth in Chapter 5, the response rate for these reports is very low. Only 69 parties have completed the report that asks this question. Of those, 35 parties left Question 1.7.5d blank, and a further four said no answer. Thus, only 30 parties provided data about confiscations of live wildlife. Their most common response was the use of public zoos and botanical gardens (27 parties). Seventeen parties used rescue centres, 11 private-approved facilities, and ten had the wildlife euthanized. Others (5) used the wildlife for training staff or gave the "specimen" to a museum (5). Presumably, this indicates that the confiscations were not only of live wildlife, but also of wildlife products. Only six parties managed to return the wildlife to the country of origin, which is the recommended course of action.

Seventeen parties indicated that they had problems with confiscated wildlife; only four answered that confiscation posed no challenges. The other parties did not answer this question (47 of the 69 participants). The challenges given were predominantly cost, but also the lack of specialized places to send wildlife as well as the problem when a seizure has lots of live wildlife; the sheer volume can be overwhelming for authorities. Only six parties provided more information about good practice related to confiscations, which included collaboration with civil society organizations and following the practical guidelines for handling seized wildlife. Confiscation is a core component of implementation, but with only six parties stating that they have returned wildlife to the country of origin, there is much more work to be done on both getting responses from parties about confiscations and also improving the outcomes from these confiscations.

Now that I have presented the data from the legislative content analysis, the Delphi iterative survey, and the interviews for the four main components of the CITES NLP, I move on to talk about other elements of implementation that I analysed, starting with captive breeding and then talking about the permit system.

Other elements of implementation

Captive breeding

In Article VII of the CITES Convention (Exemptions and Other Special Provisions Relating to Trade), the following exemption is outlined:

> 4 Specimens of an animal species included in Appendix I bred in captivity for commercial purposes, or of a plant species included in Appendix I artificially propagated for commercial purposes, shall be deemed to be specimens of species included in Appendix II.
>
> 5 Where a Management Authority of the State of export is satisfied that any specimen of an animal species was bred in captivity or any specimen of a plant species was artificially propagated, or is a part of such an animal or plant or was derived therefrom, a certificate by that Management Authority to that effect shall be accepted in lieu of any of the permits or certificates required under the provisions of Article III, IV or V.
>
> (CITES 1973: no page)

One hundred and twenty-three parties appear to mention captive breeding explicitly (60 do not have evidence of addressing the issue of captive breeding or artificial propagation). Only 15 of these 123 also explicitly include plants and artificial propagation. Table 4.8 breaks down some of the elements of the parties' legislation regarding captive breeding. The total exceeds 183 as some parties' legislation contains several elements.

Thirty-two parties' legislation seem to allow for captive breeding of CITES species but offers no further clarification as to how that might be regulated or monitored. Some of those parties that allow captive breeding, like apparently Moldova,

TABLE 4.8 Captive breeding in CITES legislation

Scientific purpose only	1	Welfare	4
Conservation purpose	11	Record sales	1
Allowed	32	Record births	1
Registered establishments	55	Permits	35

are clear that this is not for commercial purposes or trade, but specifically for scientific or conservation purposes (Republic Moldova 1995). This seems to be implied for Guyana as well, where the offspring stemming from captive-breeding operations and artificial propagation operations are the property of the state. Guyana is one of the parties that requires these operations, and wildlife ranching operations, to be licenced (Guyana 2016). Guyana's approach raises another aspect of captive breeding, that of "ranching". Several parties make the distinction between in situ and ex situ operations, where in the former the wildlife are bred in a semi-natural space, like a national park or reserve; in the latter case (ex situ), the wildlife are within a closed facility, possibly like a zoo. Other parties, like France, allow captive breeding and, presumably since Appendix I-listed species are treated as Appendix II-listed species, state that these individuals are exempt from commercial prohibitions (Legifrance 2020).

There appear to be four types of regulation and monitoring. First, there is a licensing or registration system for facilities that are engaged in captive breeding or artificial propagation. This seems to be the case for 55 parties. This corresponds to guidance by the CITES Secretariat found in Resolution Conf. 12.10 (Rev. CoP15) (CITES 2010), guidelines for the procedure to register and monitor operations that breed Appendix I animal species for commercial purposes, which recommends registration of such establishments. Permits, apparent in 35 parties' legislation, differ from registration, as they correspond specifically to the individual wildlife rather than the operation or business, which requires registration. A captive-bred (or artificially propagated) individual needs to have a permit stating that it is captive-bred. I found evidence of two further monitoring mechanisms. First, within some parties' legislation, there is the requirement to record all births at the captive breeding facility and second is the requirement to record all sales transactions. These both are probably taking place for a company's business records, but having such recording required by law improves the transparency of the industry.

A further mechanism for monitoring that was linked with registering breeding facilities was the need to be able to trace individual wildlife from birth through sale. For instance, in the Bahamas, establishments must be registered. Furthermore, wildlife must individually and permanently be marked in a manner to render alteration or modification by unauthorized persons as difficult as possible (Statute Law of the Bahamas 2016). Interestingly, in the case of the Bahamas, stock for the breeding facility may be sourced from the wild if necessary, for the health and success of the breeding, but this may only be done if doing so is not detrimental to wild populations. Similarly, Palau's legislation allows for commercial quantities of wildlife to be raised if the individuals of these species or parts thereof are identified and identifiable as having been raised under these controlled conditions. The acceptable identification must be in accordance with the regulations. Palau's approach is very comprehensive, including aquaculture, mariculture, game farming, agriculture, and horticulture (Pacific Island Legal Information Institute, no date).

One further interesting element of the captive breeding legislation of CITES parties is the explicit inclusion of welfare. This was only apparent to me in four cases where the conditions of the captivity were mentioned. For example, in Dominica, when licenses are issued for keeping wildlife in captivity, they seem to specify the conditions under which the wildlife shall be kept and disposed of (Dominica 1990). The legislation goes even further to say the license shall specify the method or methods of capture, the location where the wildlife shall be kept, and the conditions under which the wildlife shall be kept and housed. Dominica is another example where the wildlife is (remains) the property of the state.

I asked the participants of the first round of the survey whether a lack of harmonization on registering Appendix I captive-breeding facilities were a difficulty for implementing CITES. Six participants strongly agreed, 11 agreed, nine chose neither agree nor disagree, and seven disagreed. Captive breeding, then, did not rank as one of the biggest or must pressing issues in terms of CITES implementation.

Captive breeding and artificial propagation in the context of CITES are interesting in a criminological context because such enterprises provide the apparatus to potentially launder illegal wildlife, as I mentioned in the Introduction. For example, Ortiz-von Halle (2018) and Jiménez-Bustamante and Diaz Renteria (2018), among others, have found that such facilities still take wildlife from the wild and then create fraudulent documentation to make the illegal wildlife appear legal. In these examples, birds and reptiles for the pet industry constitute the wildlife who are laundered. As D'Cruz and Macdonald (2015) found in the trade of clouded leopard in Southeast Asia, the incentive to take wildlife from the wild may be even stronger when there are captive breeding facilities because it is cheaper to capture wildlife from their habitats rather than to feed and house them. "Trade exemptions for captive bred specimens may have encouraged wildlife traders to make this transition in order to meet continued consumer demand for exotic pets and wildlife tourism" (D'Cruz and Macdonald 2015: 3521). UNODC's (2020: 26) *World Wildlife Crime Report* also addresses the incidents of laundering within captive breeding operations. "This [laundering] is partly caused by conflicting and deficient legislation that reduces the effectiveness of the inspection/regulatory mechanisms. The delegation of licensing authority to sub-national administrative structures also increases the opacity of the processes and the risk of corruption".

An example of the possible supply and demand dynamics associated with captive breeding comes from the Asian arowana. The farming of the Asian arowana or dragon fish is a multimillion-dollar enterprise with complex infrastructure (Worrall 2016). When the Asian arowana was listed in CITES Appendix I, this seemed to indicate its rareness. This sparked interest in the species for the aquarium industry, and the Asian arowana is now probably the most expensive aquarium fish (Worrall 2016). In this case, the ban on trade in Asian arowana increased their demand and appears to be largely met by a substantial farming/aquaculture industry.

To me, these examples raise the possibility that captive breeding can stimulate demand by normalizing and facilitating the consumption of wildlife. By encouraging the existence of an industry that breeds wildlife (both animals and plants), are

CITES and the related organizations inadvertently increasing the trade by supporting the supply side of the trade through regulating such facilities and sending a message of the acceptability of wildlife as commodities? In some cases, at least, while potentially *not* reducing the pressure on wildlife taken from the wild, which is the purpose of captive breeding and artificial propagation in the first place. There is much more research to be done in this area, and the same may not hold true for each species of wildlife.

It is a recognized concern though, and some advice on how to tackle the possible laundering from captive breeding and artificial propagation facilities is:

> The process of issuing licenses for these types of facilities should be subject to multi-stakeholder approval involving different government departments, and anti-corruption protocols.
>
> Inspections of existing facilities should be conducted by teams focusing on a range of factors including risks for human health, sanitary compliance, environmental/ecosystem implications, integrity of the supply chain, integrity of the business model and animal welfare.
>
> Mechanisms should be developed to complement regulatory inspections with law enforcement investigations that allow for both civil and criminal justice oversight and stricter regulation.
>
> Additional focus should be put on traceability and proof of valid captive-breeding documentation throughout the supply chain.
>
> (UNODC 2020: 26)

The required import permits (for Appendix I) and the export permits (for all CITES trade) are in addition to the above permits related to captive breeding and artificial propagation and the subject of the next section.

Permits

In addition to analysing captive breeding and artificial propagation provisions in CITES legislation, I also explored how CITES parties were approaching their permit systems, since this is central to the execution of the Convention. In analysing the various permit systems, this led me to a number of parties which have stricter measures than CITES requires. In particular, both the European Union and Australia essentially treat all trade in wildlife as CITES Appendix I trade, thereby requiring import and export permits regardless of the listing of the CITES species (Australian Government 2019; European Commission 2006). Others like Fiji require more wildlife (native species in need of more protection) than that listed in the CITES appendices to be traded through the same permit system (Fiji 2017).

Overall, it appears that 102 parties' legislation mentions CITES explicitly when outlining permit requirements. A further 33 parties' legislation details a permit system of some sort, but CITES is not specifically mentioned. Another seven pieces of legislation probably are the standard CITES approach, but it was not completely clear. According to the European Commission (2018), Armenia is working with the CITES Secretariat on an electronic permit system. In fact, electronic systems and information technologies are likely to play a much more central role in not only the CITES permit system, but also reporting requirements. Decisions 18.125–18.128 encourage CITES parties to consider using the eCITES Implementation Framework (CITES 2019). Parties are asked to "consider the implementation of electronic CITES systems in a manner designed to increase transparency and efficiency of the permit issuance and control process, to prevent use of fraudulent permits, and to provide quality data for improved sustainability assessment" (CITES 2019: no page). Such implementation is supported by an off-the-shelf, low-cost system provided by the United Nations Conference on Trade and Development (UNCTAD) (CITES 2019). There is also a toolkit to support parties wanting to use electronic permits, but until all parties are able to do this, it appears paper permits also need to be used (CITES 2013).

Other approaches to permits require extra or specific permits for specific types of trade, such as Sierra Leone (1972) (among others), which mandates specific permits for trophies, or Sudan (2003), where permits are only explicitly mentioned in regard to exports of wildlife, but it is unclear about imports. Related to permits is the stance taken by parties such as Iceland, which does not allow any unauthroized imports of animals, including domestic or wild-caught (Government of Iceland 2015). Similarly, Norway used to prohibit the import and possession of live reptiles, though this has been reversed (Sollund 2019).

In regard to such stricter measures of implementation, a majority of participants of Round 1 of the Delphi iterative survey felt that these had value. For instance:

> The EU implements stricter domestic measures for all native[s] CITES II species (i.e. wolf, bear, many birds of prey, marine species)—those species have stricter protection within EU territory and are treated as CITES I species, which means that all commercial trade is prohibited with specimens taken from the wild. Many CITES parties also prohibit export of wild-taken nationally threatened species, even if they are listed in App II. Stricter domestic measures are good mechanism for preventing trade in wild-taken specimens of nationally endangered species.

Yet, one participant commented this undermines the spirit of the Convention and will result in parties not participating in CITES. As mentioned, one participant recommended that "Importing countries could respond more strongly by not allowing countries with poor implementation of or compliance with CITES to import CITES species", and this was mostly agreed with. I will discuss this more in Chapter 5, when taking a closer look at how compliance can be improved. The final element of CITES that I will discuss that is outside the NLP is animal welfare.

Animal welfare

CITES is certainly not a treaty about the welfare of the wildlife, since as a trade treaty it is essentially about the commodification of wildlife, both alive and dead. Welfare is a portion of the tension between protection versus utilization that I discussed in the Introduction, but welfare is predominantly relevant to the trade of live wildlife, both captured from the wild and being captive-bred. That is not to say CITES makes no mention of welfare. There are four relevant portions of the original text.

In Article III—Regulation of Trade in Specimens of Species Included in Appendix I—both paragraphs 3b and 5b mention care:

> a Scientific Authority of the State of import is satisfied that the proposed recipient of a living specimen is suitably equipped to house and care for it . . .
>
> a Management Authority of the State of introduction is satisfied that the proposed recipient of a living specimen is suitably equipped to house and care for it.
>
> (CITES 1973: no page)

The first covers the trade of terrestrial Appendix I wildlife and the second the introduction from the sea of Appendix I wildlife. Thus, imported live wildlife must have the proper facilities and care already in place in order to be traded.

The next welfare-related section of the Convention is Article VIII—Measures to Be Taken by the Parties. Paragraph 3 states:

> As far as possible, the Parties shall ensure that specimens shall pass through any formalities required for trade with a minimum of delay. To facilitate such passage, a Party may designate ports of exit and ports of entry at which specimens must be presented for clearance. The Parties shall ensure further that all living specimens, during any period of transit, holding or shipment, are properly cared for so as to minimize the risk of injury, damage to health or cruel treatment.
>
> (CITES 1973: no page)

Transport, then, should take place as quickly as possible and in such a way that the wildlife is not at risk from being hurt or getting ill.

Finally, in Article VII—Exemptions and Other Special Provisions Relating to Trade—the permit requirements may be waived for CITES wildlife in travelling zoos, circuses, menageries, plant exhibitions or other travelling exhibition provided that "the Management Authority is satisfied that any living specimen will be so transported and cared for as to minimize the risk of injury, damage to health or cruel treatment" (CITES 1973: no page).

Essentially then, the welfare-related portions of CITES pertain to the conditions during transport, the responsibility of which falls on the exporter, and the conditions at the destination, which falls on the importer. One Resolution—Conf. 10.21 (Rev. CoP16) Transport of live specimens—provides guidelines as to how transport should be conducted. CITES advises that the Live Animal Regulations (LAR) of the International Air Transport Association (IATA) should be used when transporting live wildlife by plane (CITES 1995 [2013]). To access the LAR-IATA regulations for transporting live wildlife, the party would have to pay USD 300 to purchase a copy of the regulations (IATA 2020). I suggest this cost is prohibitive for many parties. Judging from the checklist for live animal transport, the focus is on the container the wildlife is held in and that the paperwork is in order. Food, water, and bedding may be "not applicable". For transport not by air, Conf. 10.21 recommends parties look at the *CITES guidelines for the non-air transport of live animals and plants*. As a Resolution, again, these are only recommendations and are not mandatory for the parties to utilize.

As part of my legislative content analysis, I attempted to assess whether or not parties' legislation contained language that covered the conditions of transport as well as the requirement for adequate housing and care at the destination. I also made note of mention of the IATA regulations and whether the parties' legislation contained other welfare provisions. Nearly 57 per cent of the parties' legislation has no apparent incorporation of any welfare provision (104 parties). A further three have statements that are unclear as to the welfare elements that are being included. For instance, there is legislation that says "appropriate" care for the individual. The legislation of 36 of the parties contains welfare provisions, but not in the language of CITES. The remaining 40 parties have some features of the required welfare provisions of CITES. Twenty-three of these contain both the export conditions and the housing and care at the destination. Sixteen of these 40 parties' legislation only seems to contain reference to the export conditions—the minimization of risk of injury, damage to health or cruel treatment during transport. Two pieces of legislation seem to only incorporate the check of the housing and care of the facility where the wildlife will be held. Nine parties specifically mention the use of IATA's Live Animal Regulations. Finally, five parties' legislation makes specific reference to a separate welfare act.

Of the pieces of legislation that do not specifically utilize the CITES language in relation to welfare, some are quite progressive in their regard for wildlife. An example of the 24 pieces of legislation that I categorized as "other" is South Korea's Wildlife Protection and Management Act (2014):

> No one shall treat any of the following wild animals with cruelty without a justifiable ground:
>
> 1 Killing wild animals in a manner disgusting to other persons, such as beating or burning alive;
> 2 Causing pain to the wild animals or inflicting any wound on them after capturing and caging them;

3 Installing any device in the body of any wild animal to collect blood, gallbladder, internal organs, or parts of the living body of wild animals or collecting them;
4 Killing wild animals cruelly, such as strangulation or the use of toxic chemicals.

The legislation clearly targets some practices associated with Traditional Asian Medicines (milking bears for gall) and cruel forms of killing.

The above data support previous research that engagement with welfare is very limited by CITES and its parties (Maher and Wyatt 2017; Sollund 2019; Wyatt 2013; Wyatt et al. Under review b). Even though the welfare conditions for live transport have been in a Resolution since 1995, a minority of legislation by parties actually contain elements of welfare, and these are confined to transportation and somewhat to housing, but only of individuals being imported. In my limited experience of the CoP, welfare is a tangential discussion. The one-sided event on the ethics of wildlife trade was cancelled with no explanation from the organizer. The IFAW and the IUCN, however, have produced guidelines on the confiscation of live wildlife (IFAW and IUCN 2019). These are freely available to all CITES stakeholders and are a valuable resource to improve this particular aspect of CITES wildlife trade. As discussed more in Chapter 7, the lack of a wider discussion of non-human animal welfare is potentially problematic, as non-human animal welfare and rights underpin many of the discussions at CITES and are only going to increase.

Now, I move on to discuss what the literature and data have indicated as the limitations and difficulties faced by parties trying to implement CITES in general and in terms of the four components of the NLP.

Implementation limitations and difficulties

Hopefully, it is clear that there is a very mixed picture when it comes to how and which parties implement the four main components of the CITES convention as assessed under the NLP as well as other elements (i.e. captive breeding/artificial propagation and permits). This section explores reasons for why much of the implementation is not at a level that might be expected, even after 45 years since CITES' inception. Reeve (2002) and Wijnstekers (2018), among others, note that there are a number of difficulties for parties to fully implement CITES. The difficulties cited typically revolve around lack of human and financial resources. Since 2015, when the CITES biennial reports became implementation reports, the questions asked of the parties for these reports centre on implementation challenges. The questions that I asked in my Delphi iterative survey draw largely on the questions in these biennial reports. The intention was to develop some consensus from CITES experts around specific actions that might be taken to address the shortcomings in implementation.

I grouped my questions into the two categories found in the CITES biennial reports. I suggest difficulties (see Table 4.9) are actions required by parties, which are possible or doable but hard to complete. Constraints (see Table 4.10) are outside forces and/or conditions that are, for the most part, out of the Management and Scientific Authorities' control. It might appear that there is some overlap here. For instance, the lack of finances and human resources is listed as difficulties, and insufficient resources are listed as a constraint. To me, these are distinctive because there are parties which could better fund and provide staff, which do not (a difficulty), and there are also parties, which do not have the means to provide the finances or the personnel to implement CITES (a constraint).

Lack of finances was thought to be the biggest difficulty followed by lack of human resources. A majority of participants agreed (10 strongly agreed and 16 agreed) the lack of trade data from some parties made implementation of CITES

TABLE 4.9 Difficulties in implementing CITES

Improvement or Alternative	*Strongly Agree*	*Agree*	*Neither*	*Disagree*	*Strongly Disagree*
Tight deadlines for reporting	1	11	6	13	1
Tight deadlines for implementing new resolutions	3	13	5	10	1
Data asked for are difficult to collect	6	9	6	10	0
Hard-to-implement resolutions such as on musical instruments, which generate added work and seizures	4	18	3	5	1
Lack of data from some members makes implementation difficult	10	16	4	2	1
Lack of finances	14	17	0	1	0
Incomplete records with no scientific names	5	13	8	4	1
Lack of human resources	16	13	0	3	0
Awareness of relevant organizations is limited	8	12	6	6	0
Not all resolutions and decisions can be transposed into community law as needed	3	17	6	4	1
The regulation of illegal trade is compromised by concerns that it could adversely affect the legal trade, and thus the material interests of many member countries	5	10	8	5	2
Appendix listings and resolutions are based on political/financial interests rather than scientific or other evidence	5	11	4	9	3

TABLE 4.10 Constraints to implementing CITES

Improvement or alternative	*Strongly Agree*	*Agree*	*Neither*	*Disagree*	*Strongly Disagree*
Coordinating multiple agencies involved in implementation	8	20	2	1	0
Overly complicated import-export procedures	4	12	2	12	2
International pressure	3	9	8	8	3
Illegal trade	11	12	6	3	0
Insufficient resources (material, financial, human, technical)	14	17	0	1	0
Weak capacity (human and material) of authorities	17	14	0	1	0
Legislative development	3	18	7	4	0
Corruption	15	10	5	0	2

difficult. Participants were split on whether the logistical requirements of CITES were difficulties. So, the deadlines for reporting, the deadlines for implementing new resolutions, and the collection of the data were, for some, viewed as a difficulty, but not viewed so for others. The tension between protection and use was mentioned by one participant as a difficulty in implementation; they felt that there were too many abolitionists with their feet dug into the ground on sustainable regulated use of wildlife.

The lack of finances and human resources—insufficient resources—was also one of the most agreed-upon constraints (14 strongly agreed and 17 agreed). Even more than for resources, participants strongly agreed that corruption (15 participants) and weak capacity of authorities (17 participants) are constraints. Participants were divided as to whether or not international pressure and overly complicated import-export procedures (the permit system) are a constraint on CITES implementation. One participant suggested that a constraint not listed is the pressure to continue trading from traders.

My survey participants and interviewees as well as the parties themselves, as is evident from the biennial (implementation) reports, have a clear sense of the many elements of implementation that could be improved. In the revised version of the biennial report from 2015, parties are asked several questions about what they are doing to improve their implementation and what their priorities are in this regard. The answers provide further useful insight into the ongoing challenge of improving CITES implementation.

Parties' efforts for implementation

Question 2.2.2a of the biennial (implementation) reports post-2015 asks: "Have any of the following activities been undertaken during the period covered in this report to enhance the effectiveness of CITES implementation at the national level"? (CITES 2018: 14). There are tick box answers for: hiring of more staff, development of implementation tools, purchase of technical equipment for implementation, monitoring or enforcement, and other (please specify). I was particularly interested in whether parties had made efforts in relation to development of implementation tools. Thirty-five parties left this question blank; five did not provide an answer for this portion of Question 2.2.2a, and eight said the question was not applicable. Nine parties said "no", they had not developed any implementation tools, and 12 said "yes", they had. Unfortunately, no further detail is provided as to what those tools are, at least in this section. (I discuss in much more detail, the poor reporting rate of the parties for these reports in Chapter 5, when I talk about compliance).

The Secretariat is interested in more than just who the CITES authorities are in the parties (one of the four components of implementation assessed for categorization, as discussed thoroughly when looking at the NLP), but asks in Question 2.2.2c: "Have you been able to use international development funding assistance to increase the level of implementation of your Management, Scientific and Enforcement authorities"? (CITES 2018: 2015), with a yes, no, or not applicable tick box for each authority. For the Management Authority, five parties said yes, and ten parties each said no or not applicable. The other 44 parties completing the report left this question unanswered. In terms of the Scientific Authority, four parties responded that they had received such assistance. Nine answered not applicable, and 12 answered no, with the remaining 44 parties again leaving the question blank. Interestingly, the Secretariat asks about Enforcement Authorities though they are not required. Six parties answered "yes", nine "no", and ten "not applicable". Again, 44 parties did not answer this particular question.

The report asks in Question 2.2.2d: "What is the respective level of priority for enhancing the effectiveness of CITES implementation at the national level through the following activities"? (CITES 2018: 15). The topics of concern reflect Questions 2.2.2a: hiring of more staff, development of implementation tools, purchase of technical equipment for implementation, monitoring or enforcement, and other (please specify), but also includes e-permitting. Each topic is followed by a tick box for high, medium, low, and not a priority (see Table 4.11). However, as no party chose "not a priority", the table with the results has a column for "not answered" instead. Hiring more staff is the highest priority for most of the parties that answered. This is followed by e-permitting, and then development of implementation tools and purchase of technical equipment are of equal priority (11 parties each). It would be interesting to know in which authority more staff are thought to be needed. This would give a better sense of the parties' efforts to improve CITES. For instance, if more staff are needed in the Enforcement Authority, this is technically not an implementation priority, but an enforcement effort.

The move to an implementation report is welcomed, as it supposedly provides a more detailed level of information about the state of implementation. Yet, much of the content almost seems too far advanced or actually not directly related to

TABLE 4.11 Parties' priorities for implementation

Issue	*PRIORITY*			
	High	*Medium*	*Low*	*Not Answered*
Hire more staff	16	8	5	3
Development of implementation tools	11	15	5	1
Purchase of technical equipment for implementation, monitoring or enforcement	11	13	8	0
E-permitting	13	13	5	1
Other	0	0	0	0

implementation, but rather related to enforcement. Category 3 parties with legislation that does not meet the basic implementation requirements may not be worrying about developing tools to implement the Convention, when they do not have all the required authorities or do not prohibit violations. Some of the questions also appear to need further explanation or examples. For instance, it is not clear what a development tool might be or what equipment (besides e-permitting) would monitor and evaluate implementation. Perhaps further amendments could be made to this report to first link the questions more directly to the NLP and then ask more fine-grained questions than are currently asked. With that in mind, in the next section, I explore what improvements could be made to the NLP itself, which, as discussed earlier, is intended to assist with implementation.

The four components of the NLP

Considering that the NLP was (is) designed to improve implementation of the Convention and that it has been active since 2002, I asked participants about its effectiveness and how it might be improved. In Round 1 of the Delphi iterative survey, Question 16 asked "Does the listing of a country in Category 1 (fully implemented), 2 (partly implemented), and 3 (have not implemented) by the NLP accurately capture whether a country has implemented CITES"? The responses were: Strongly Agree 1, Agree 10, Neither 10, Disagree 8, and Strongly Disagree 3. Seemingly, participants were divided as to whether the categories of the NLP truly reflect the implementation of the Convention into national legislation.

Other studies have indicated that the categories may not accurately reflect a party's legislation. For example, a wildlife trade policy review of Madagascar finds several problems with implementation (INBA 2009). This comprehensive review states that at times Madagascar has been found:

> exceeding quotas, illegal trafficking, setting quotas which are not scientifically justified. . . . Measures adopted by the international community were

> necessary to help Madagascar limit its rate of export during some periods: the reduction of the number of chameleon and *Phelsuma* species which could be exported from 1994; the repeated refusal of the European community to allow the import of several species of fauna and flora (*Uroplatus, Mantella, Pachypodium*, . . .). This reflects a lack of content and a failure to implement the policy. There are no accompanying measures to correctly apply the provisions adopted and officials lack the will to implement them.
>
> (INBA 2009: 62–63)

Yet, despite these shortcomings, including a need to complete Reviews of Significant Trade, which are a requirement that I discuss in Chapter 5, Madagascar's legislation is Category 1.

From my legislative content analysis, I would lean towards disagreeing that the NLP accurately reflects whether or not a party's legislation implements CITES. There are several parties' legislation where it is unclear to me why they are not categorized higher and there are other parties' legislation where is it unclear to me why they are not categorized lower. I wonder about 50 parties' categorizations:

TABLE 4.12 Differing categorizations of implementation between the NLP and the legislative content analysis

	NLP Category	*My Analysis Category*	*Number of Parties' Legislation*
Why not a higher NLP category?	3	2	18
	2	1	11
	3	1	2
Why not a lower NLP category?	1	2	19
	2	3	0
	1	3	0

The "My Analysis Category" column contains my assessment of what the NLP category would be based on the legislative content analysis. To be Category 1, all four components are present; one, two, or three components is Category 2; and no components is Category 3. All of the 20 parties that are 3:2 and 3:1 (NLP: My Analysis) have some amount of implementation. In the case of the two 3:1 parties—Dominica and Nepal—my analysis indicated that there was evidence of all four components (Dominica 1990; Nepal Law Commission 2018). On the reverse side, I found 19 parties' legislation that the NLP categorized as 1, but for which during my analysis I could not identify all four of the NLP components. Twelve of those 19 are missing information about confiscations, but other than that, there is no clear trend as to why these parties seem to be categorized higher than apparently warranted. Part of the problem is that the NLP does not break down which components are present or absent when they categorize a party. Nor is there further information on whether one component is weighed more than

TABLE 4.13 What mechanism other than NLP categories might be a better indicator of implementation? Or what improvements could be made to the NLP?

Improvement or Alternative	*Strongly Agree*	*Agree*	*Neither*	*Disagree*	*Strongly Disagree*
More detailed breakdown of the four components, i.e. which component is lacking and how	10	18	3	1	0
Metrics related to annual and biennial report submission linked to categories	6	15	10	0	0
Add new components, i.e. standard level of penalty	6	18	5	3	0
More visibility on CITES website and in the media about a country's category	6	17	4	3	1
Make the NLP a project independent of CITES	6	3	16	6	1
Metrics related to delistings, uplistings, and extinctions linked to categories	1	11	15	4	0
Use victimization devices to measure implementation	4	11	6	6	0

another component during categorization. I explored this lack of transparency in the first Round of the Delphi iterative survey, along with asking participants what other components might be better to indicate proper implementation of CITES.

Participants agreed (10 strongly agreed; 18 agreed) with my observation that it would be useful for the NLP to break down the lacking components and how they are lacking during the categorization. There were mostly non-committal responses to making the NLP an independent project not run by CITES (16 neither agree nor disagree) and to adding a metric related to down-listings, up-listings, and extinctions (15 neither agree nor disagree). The latter was asked since it is another way of judging whether CITES is effective (i.e. protecting wildlife in a sustainable way since they are not coming under more pressure—up-listed or extinct).

The last suggestion that I made was to use victimization devices to measure implementation. Again, the impetus behind this question was that there are other ways of judging effective implementation. In this case, I explained victimization device measures "may include a combination of statistical records of violations (illegal trade) and testimonials (from human communities impacted by overexploitation and illegal trade and potentially from people representing the wildlife)". Whereas illegal trade reports have fairly recently been re-introduced by the CITES Secretariat (though this still does not form part of the NLP components), there, as far as I am aware of, has never been direct engagement with communities living with and near CITES-listed wildlife to gain their perspective on implementation of and compliance with CITES.

In regard to improving the NLP, I also asked participants to make their own suggestions. These proved to be quite insightful. In particular, one participant recommended a more hands-on approach to the NLP, rather than it being largely a desk-based exercise: "In-situ assessment and verification missions by Secretariat". Another participant suggested, "Successful prosecutions is a key metric that should be considered. Also—penalties allowed. Also, should evaluate whether the legislation covers ALL CITES-listed species (not only native species; some countries don't cover marine species)". The latter point, as previously mentioned, is a particularly important one that was supported by my analysis, which clearly showed the legislation was centred on terrestrial animals with sometimes no inclusion of plants or marine species (Wyatt et al. Under review a).

Finally, another participant said,

> Having correctly worded legislation does not mean that said legislation is being correctly and adequately enforced. Issues to be considered include a measure of the resources invested in enforcement, the success of investigations/prosecutions, the level of government corruption, the level of public support, etc. There is a saying that "laws without enforcement are just suggestions". That tends to be quite true.

As indicated, though, at the beginning of this chapter, there is no requirement for an Enforcement Authority, and obviously enforcement (arrests and prosecutions) of CITES is not a current metric of the NLP or (yet) in the annual or biennial reports.

Improvements to implementation

Regardless of the possible flaws of the NLP, it is a valuable tool to have some oversight of implementation of CITES requirements into the parties' national legislation. As will be evident, there is clearly scope for the implementation to be improved, and I suggest the following in regard to the four components of the NLP, before moving on to some concluding remarks on implementation.

Authorities

- Concentrated effort to get all members to have a Scientific Authority.
- Consider adding an Enforcement Authority to the Convention.

Prohibition

- Ensure all members prohibit violations of the Convention in native and non-native species.
- Ensure all members prohibit violations of the Convention in terrestrial and marine fauna and flora.

Penalties

- Consider a Resolution or Decision to set penalties, possibly in line with the UN guidelines that 4+ years indicates a serious crime.
- Audit that penalties are being given and to what levels and extent (this would also capture the number of violations, arrests, and prosecutions).

Confiscations

- Support parties to widen confiscations to include assets and proceeds of crime.
- Support parties to develop robust evidence procedures and storage facilities.
- Wider dissemination of the IFAW and IUCN guidelines on the confiscation of live wildlife (IFAW and IUCN 2019).

Conclusion

Obviously then, implementation of CITES is multifaceted, multilayered, and complex. Full implementation is thwarted by difficulties and constraints even in parties with high levels of political will. But even for parties that appear to have fully implemented the Convention, there is doubt that this is the case. My analysis indicates several parties in Category 1 that may well have gaps in their implementation. Since the NLP does not provide details in relation to the exact implementation of the four components, it is hard to be sure. On the flip side, there are also parties that appear to be doing better than how the NLP has categorized them. Still, overall, even if my categorizations more closely capture the state of implementation of the CITES parties, less than half would be meeting their obligations.

Interestingly, the idea of powerful interests of those profiting from wildlife trade as a barrier to implementation was barely mentioned. At least the experts I spoke to, in general, seem to believe the overriding issues in regard to implementation are resources (financial and human), weak capacity, and corruption. Corruption may include powerful interests, but further research is needed to unpack what is considered corruption in the context of CITES implementation. There are concrete steps that can be taken to improve implementation, which should be considered. The same is true for improving compliance, which I detail in the next chapter.

References

Arroyo-Quiroz, I. and Wyatt, T. 2019. Wildlife trafficking between the European Union and Mexico. *International Journal for Crime, Justice and Social Democracy*. 8 (3): 23–37. https://doi.org/10.5204/ijcjsd.v8i3.1243.

Australian Government. 2019. Environment protection and biodiversity act 1999. Available at: www.legislation.gov.au/Details/C2019C00275. Accessed 10 August 2020.

Bowman, M., Davies, P. and Redgwell, C. 2010. *Lyster's International Wildlife Law*. Cambridge: Cambridge University Press.

Convention on the International Trade in Endangered Species of Wild Fauna and Flora (CITES). 2019.18.125–18.128 Electronic systems and information technologies. Available at: www.cites.org/eng/dec/valid17/82172. Accessed 11 August 2020.

———. 2018. Implementation report 2015–2017 Austria. Available at: https://cites.org/sites/default/files/reports/15-17Austria.pdf. Accessed 11 December 2020.
———. 2013. CITES electronic permits toolkit version 2.0. Available at: https://cites.org/sites/default/files/eng/prog/e/cites_e-toolkit_v2.pdf. Accessed 11 August 2020.
———. 2010. Guidelines for a procedure to register and monitor operations that breed Appendix-I animal species for commercial purposes. Resolution Conf. 12.10 (Rev. CoP15). Available at: www.cites.org/sites/default/files/document/E-Res-12-10-R15.pdf. Accessed 10 August 2020.
———. 1995 [2013]. Resolution Conf. 10.21 (Rev. CoP16). Transport of live specimens. Available at: https://cites.org/sites/default/files/document/E-Res-10-21-R16.pdf. Accessed 11 December 2020.
———. 1973. Convention on international trade in endangered species of wild fauna and flora. Available at: www.cites.org/eng/disc/text.php#IX. Accessed 30 July 2020.
D'Cruz, N. and Macdonald, D. 2015. Clouded in mystery: The global trade in clouded leopards. *Biodiversity Conservation*. 24: 3505–3526.
DLA Piper. 2015. *Empty Threat 2015: Does the Law Combat Illegal Wildlife Trade? A Review of Legislative and Judicial Approaches in 15 Jurisdictions*. Washington, DC: DLA Piper and WWF.
Dominica. 1990. Forestry and wildlife act. Chapter 60:02. Available at: www.dominica.gov.dm/laws/chapters/chap60-02.pdf. Accessed 7 August 2020.
European Commission. 2006. Commission Regulation (EC) 865/2006 laying down detailed rules concerning the implementation of Council Regulation (EC) No 338/97 on the protection of species of wild fauna and flora by regulating trade therein. Available at: www.legislation.gov.uk/eur/2006/865/adopted. Accessed 10 August 2020.
———. 2018. Joint staff working document: The EU special incentive arrangement for sustainable development and good governance ('GSP+') assessment of Armenia covering the period 2016–2017. Available at: http://trade.ec.europa.eu/doclib/docs/2018/january/tradoc_156537.pdf. Accessed 11 August 2020.
Felbab-Brown, V. 2018. *The Extinction Market: Wildlife Trafficking and How to Counter it. International Affairs*. London: Hurst & Company.
Fiji. 2017. Act No 10 of 2017—An Act to Amend the Endangered and Protected Species Act 2002. Available at: http://extwprlegs1.fao.org/docs/pdf/fij171675.pdf. Accessed 11 August 2020.
Government Gazette of the Republic of Namibia. 2008. No. 4190. Available at: https://cites.org/sites/default/files/projects/NLP/NamibiaControlledWildlifeProductsandTradeAct9.2008.pdf. Accessed 28 July 2020.
Government of Iceland. 2015. Importation of animals acts 1990 No. 54, 16 March. Available at: www.government.is/library/04-Legislation/Act%20no%2054_1990-Importation%20of%20Animals%20Act-mai-2015.pdf. Accessed 11 August 2020.
Guyana. 2016. Wildlife conservation and management bill 2016. Available at: http://dev.ultimate-dimensions.net/nredev/wp-content/uploads/2016/06/WILDLIFE-CONSERVATION-AND-MANAGEMENT-BILL-2016.pdf. Accessed 10 August 2020.
INBA. 2009. National wildlife trade policy review. Madagascar. Available at: www.inba.info_car.pdf. Accessed 14 December 2020.
International Air Transport Association (IATA). 2020. 2021 live animal regulations. Available at: https://store.iata.org/IEC_ProductDetails?id=9105-47&_ga=2.235785729.1262246004.1607673992-118588979.1607673992. Accessed 11 December 2020.
International Fund for Animal Welfare (IFAW) and International Union for the Conservation of Nature (IUCN). 2019. *Guidelines for the Confiscation of Live Wildlife*. Gland, Switzerland: IUCN.
Jiménez-Bustamante, D. and Díaz Rentería, L. P. 2018. Laundering of illegal wild fauna in Mexico: Case study of a pair of desert monitors *Varanus griseus griseus* (Daudin, 1803).

In Arroyo-Quiroz, I. and Wyatt, T. (eds.). *Green Crime in Mexico: A Collection of Case Studies*. Palgrave Studies in Green Criminology. London: Palgrave Macmillan: 127–136.

Legifrance. 2020. Arrêté du 30 juin 1998 fixant les modalités d'application de la convention sur le commerce international des espèces de faune et de flore sauvages menacées d'extinction et des règlements (CE) n° 338/97 du Conseil européen et (CE) n° 939/97 de la Commission européenne. Available at: www.legifrance.gouv.fr/affichTexte.do?cidTexte=LEGITEXT000005626370. Accessed 5 August 2020.

Maher, J. and Wyatt, T. 2017. International trade in animals and animal parts. In Beirne, P., Maher, J. and Pierpont, H. (eds.). *International Handbook on Animal Abuse Studies*. Basingstoke: Palgrave Macmillan: 223–247.

Marceau, J. 2019. *Beyond Cages: Animal Law and Criminal Punishment*. Cambridge: Cambridge University Press.

Nepal Law Commission. 2018. An act to regulate and control international trade in endangered wild fauna and flora. Available at: www.lawcommission.gov.np/en/archives/category/documents/prevailing-law/statutes-acts/an-act-to-regulate-and-control-international-trade-in-endangered-wild-fauna-and-flora. Accessed 13 August 2020.

Ortiz-von Halle, B. 2018. *Bird's Eye View: Lessons from 50 Years of Bird Trade Regulation & Conservation in Amazon Countries*. Cambridge: TRAFFIC.

Pacific Island Legal Information Institute. No date. Environmental protection—title 24—chapter 10. Available at: www.paclii.org/pw/legis/consol_act/ept24335/. Accessed 7 August 2020.

Reeve, R. 2002. *Policing International Trade in Endangered Species: The CITES Treaty and Compliance*. London: Earthscan.

Republic Moldova. 1995. Regnului animal. Available at: http://extwprlegs1.fao.org/docs/pdf/mol74856.pdf. Accessed 5 August 2020.

Sierra Leone. 1972. The wild life conservation act 1972. Available at: http://extwprlegs1.fao.org/docs/pdf/sie41659.pdf. Accessed 11 August 2020.

Sollund, R. 2019. *The Crimes of Wildlife Trafficking: Issues of Justice, Legality and Morality*. London: Routledge.

South Korea. 2014. Wildlife protection and management act. Available at: http://extwprlegs1.fao.org/docs/pdf/kor163043.pdf. Accessed 11 December 2020.

Statute of the Bahamas. 2016. Chapter 250A Wildlife conservation and trade. Available at: http://extwprlegs1.fao.org/docs/pdf/bha78747.pdf. Accessed 7 August 2020.

Sudan. 2003. Laws of the new Sudan: Wildlife conservation and national parks act, 2003. Available at: http://extwprlegs1.fao.org/docs/pdf/ssd152460.pdf. Accessed 11 August 2020.

United Nations Food and Agricultural Organisation (UNFAO). 2012. Wildlife (Conservation and security) act, 2012. Available at: http://extwprlegs1.fao.org/docs/pdf/bgd165019.pdf. Accessed 31 July 2020.

United Nations Office on Drugs and Crime (UNODC). 2020. *World Wildlife Crime Report: Trafficking in Protected Species*. Vienna: UNODC.

———. 2016. *World Wildlife Crime Report: Trafficking in Protected Species*. Vienna: UNODC.

———. 2008. On rules and procedures for international trade in endangered species of fauna and flora of the wild. Available at: www.unodc.org/res/cld/document/law-no-9867-on-rules-and-procedures-governing-international-trade-of-endangered-species-of-wild-fauna-and-flora-english_html/Law_No._9867_English.pdf. Accessed 31 July 2020.

Waterland, S., Vaughan, J., Lyman, E. and Jurisic, I. 2015. Illegal wildlife trade review, Malawi. Available at: www.lilongwewildlife.org/wp-content/uploads/IWT-Review-Malawi.pdf. Accessed 30 July 2020.

Wijnstekers, W. 2018. *The Evolution of CITES: A Reference to the Convention on International Trade in Endangered Species of Wild Fauna and Flora, 11th Edition*. International Council for Game and Wildlife Conservation. Available at: www.cic-wildlife.org/wp-content/uploads/2018/05/the-evolution-of-cites-2018-11th-editionpart-1.pdf. Accessed 27 November 2019.

Worrall, S. 2016. Meet the world' most expensive pet fish. *National Geographic*. Available at: www.nationalgeographic.com/news/2016/07/dragon-fish-most-expensive-arowana-emily-voigt/?cmpid=org=ngp::mc=crm-email::src=ngp::cmp=editorial::add=SpecialEdition_Escape_20200805&rid=720BDCA3E3D4E41A15BE807B71F01F25. Accessed 7 August 2020.

Wyatt, T. 2013. *Wildlife Trafficking: A Deconstruction of the Crime, the Victims and the Offenders*. London: Palgrave Macmillan.

Wyatt, T. Friedman, K. and Hutchinson, A. Under review a. Are fish wild? *Animal Law Review*.

Wyatt, T., Maher, L., Allen, D., Bruschi, N., Clarke, N. and Rook, D. Under review b. The welfare of the wildlife: How the legal and illegal trade debates ignore the wildlife. *Crime, Law and Social Change*.

Wyatt, T., Van Uhm, D. and Nurse, A. 2020. Differentiating criminal networks in the illegal wildlife trade: organized, corporate and disorganized crime. *Trends in Organised Crime*. https://doi.org/10.1007/s12117-020-09385-9.

5

CITES COMPLIANCE

The big picture

"CITES Compliance: the big picture" is the focus of Chapter 5. This chapter analyses the ways in which *parties* have and have not complied with CITES requirements. As I delineated in Chapter 2, this type of compliance is separate and different from the Convention being complied with by individual people or companies trading wildlife. To comply with the requirements of the Convention, parties must not only implement the Convention provisions (the subject of Chapter 4), but also meet the reporting and review requirements and update national legislation to reflect the changes that take place at the CoP. In the case of CITES in terms of reporting, as mentioned, this means an annual report and a biennial (implementation) report. Review requirements are "responding to specific recommendations of the Standing Committee concerning the implementation of Article IV of the Convention in the context of the Review of Significant Trade" (CITES 2019: no page). In terms of legislative updates, for the most part, this relates to additions of species to the CITES appendices or movement of species on the appendices. In this chapter, I will detail the parties' compliance with both annual and biennial (actually triennial) reports, with undertaking Review of Significant Trade, and their compliance with making changes and updates. I will then discuss the mechanisms in place for addressing non-compliance by the parties. Again, the data underpinning my analysis are taken from CITES sources and primary data from the Delphi iterative survey and expert interviews as well as the literature.

Annual reporting

As mentioned, the CITES Secretariat and Standing Committee monitor one part of compliance with the Convention in terms of annual and biennial reporting. In this section, my focus is on annual reporting. Article VIII—Measures to Be Taken by Parties—sets out the elements of annual reporting:

> 7 Each Party shall prepare periodic reports on its implementation of the present Convention and shall transmit to the Secretariat:
>
> (a) an annual report containing a summary of the information specified in sub-paragraph (b) of paragraph 6 of this Article.
>
> (CITES 1973: no page)

Subparagraph (b) of paragraph 6 of Article VIII states,

> (b) the number and type of permits and certificates granted; the States with which such trade occurred; the numbers or quantities and types of specimens, names of species as included in Appendices I, II and III and, where applicable, the size and sex of the specimens in question.
>
> (CITES 1973: no page)

Parties, as dictated in Article VIII, paragraph 6, subparagraph (a), are also supposed to keep records of the names and addresses of exporters and importers (CITES 1973). These are not required to be shared, for obvious reasons of privacy, but it is unclear to me whether or not there is an auditing process to ensure this record-keeping is taking place. There seems to be no checking either about the levels of accuracy or standards for the records for exporters and importers though there are guidelines (see later). The final part of Article VIII states that the annual (and biennial) reports shall be available to the public where this is permitted by the parties' legislation (CITES 1973). Presumably, this requirement is met through the CITES trade database run by the UN Environment Programme's Wildlife Conservation Monitoring Centre (UNEP-WCMC), which is available on the CITES website. The trade database is a collation of all the parties' annual report data into downloadable information.

The original text of the Convention does not specify when annual reports should be submitted to the Secretariat. This is clarified in Resolution Conf. 11.17 (Rev. CoP18) "National Reports". As mentioned, Resolutions are not obligatory, which is clear from the language used in the text:

> 1 URGES all Parties to submit their annual reports required under the provisions of Article VIII, paragraph 7 (a), by 31 October following the year for which they are due and in accordance with the most recent version of the *Guidelines for the preparation and submission of CITES annual reports* distributed by the Secretariat, as may be amended with the concurrence of the Standing Committee
>
> (CITES 2000 [2019]: 1)

That is not to say that the Secretariat does not raise the issue of annual reports not being submitted within the urged deadline. Resolution Conf. 11.17 (Rev. CoP18) (2000 [2019]: 2) goes on to state:

> 13 DECIDES that:
>
> a) failure to submit an annual report by 31 October of the year following the year for which the report was due constitutes a major problem with the implementation of the Convention, which the Secretariat shall refer to the Standing Committee for a solution in accordance with Resolution Conf. 11.3 (Rev. CoP18) on *Compliance and enforcement*; and
>
> b) the Secretariat may approve a valid request from a Party for a reasonable extension of time to the 31 October deadline for the submission of national reports provided the Party submits to the Secretariat a written request, containing adequate justification, before that deadline.

Thus, the Secretariat is understanding of the challenges in reporting. And, as mentioned earlier, a former CITES Secretary General has said that it is particularly difficult for developing countries to comply with the Convention (Wijnstekers 2018). The recognition of the difficulty to comply is further evidenced in that in this Resolution parties are urged to seek assistance from the Secretariat if they are struggling (CITES 2000 [2019]). As one Round 1 Delphi iterative survey participant stated, "The CITES compliance system allows a staged process from capacity building to remedial measures, such as trade suspensions". Furthermore, it is not until 3 years of non-compliance that more serious steps to gain compliance are recommended:

> 14 INSTRUCTS the Standing Committee to determine, on the basis of reports presented by the Secretariat, which Parties have failed, for three consecutive years and without having provided adequate justification, to provide the annual reports required under Article VIII, paragraph 7 (a), of the Convention within the deadline (or any extended deadline) provided in the present Resolution;
>
> 15 RECOMMENDS that Parties not authorize trade in specimens of CITES-listed species with any Party that the Standing Committee has determined has failed, for three consecutive years and without having provided adequate justification, to provide the annual reports required under Article VIII, paragraph 7 (a), of the Convention within the deadline (or any extended deadline) provided in the present Resolution.
>
> (CITES 2000 [2019]: 2)

The support for parties to become compliant with the Convention is not only from the Secretariat. Decisions have been drafted that encourage other parties to provide financial or technical support to parties subject to compliance measures in order to strengthen the institutional capacity of the non-compliant party (IISD 2019).

Yet, as Bowman et al. (2010: 522) state,

> "the information derived from national reports reveals a limited picture of the true extent of international trade in wildlife. Too often those reports that are actually submitted are not of a high quality and can be incomplete in that only trade in certain species is covered. This is often due to the fact that various institutions at the national level have responsibilities in the area of wildlife trade but only some contribute to the national annual report".

The insight that key stakeholders do not contribute to the data that is collated for the annual report is worrisome. Somehow, then, an auditing system for the quality and completeness of the data provided by the parties is needed.

Even with a fairly flexible and supportive approach to annual reporting, as stated in Chapter 1 when justifying why this research is important, there are a number of parties, which are still not complying with the reporting requirements. In terms of annual reports, Afghanistan, Djibouti, and Grenada are, at the time of writing, under temporary trade suspensions of all species until they comply (CITES 2019). As I stated previously and is acknowledged by the Secretariat in Resolution Conf. 11.17 (2000 [2019]), annual reporting is critical to knowing the status of species and any concerns with declining populations or illegal trade. The annual reports and the resulting UNEP-WCMC database of trade in CITES species underpin the Review of Significant Trade that forms the foundation for much of the trade in Appendix II-listed species.

Review of significant trade

As described earlier, trade suspensions for some parties are due to their consistent failure to submit annual reports. This is also the case for failure to undertake Reviews of Significant Trade. Such reviews are not specifically prescribed in the original Convention text. The process has arisen out of Article IV—Regulation of Trade in Specimens of Species Included in Appendix II—where trade in Appendix II-listed species is allowed with an export permit. To issue an export permit, the exporting party's Scientific Authority needs to advise that such export will not be detrimental to the survival of the species (CITES 1973). The exporting party's Management Authority needs to be satisfied "the specimen was not obtained in contravention of the laws of that State for the protection of fauna and flora" and "that any living specimen will be so prepared and shipped as to minimize the risk of injury, damage to health or cruel treatment" (CITES 1973). Furthermore, the Scientific Authority can determine that a limited number of "specimens" should be exported "to maintain that species throughout its range at a level consistent with its role in the ecosystems in which it occurs and well above the level at which that species might become eligible for inclusion in Appendix I" (CITES 1973: no page). They inform the Management Authority on the measures that need to be taken to limit exports.

Several attempts to improve the monitoring of Appendix II trade appear to have been unsuccessful, and are referred to in Resolution Conf. 12.8 (Rev. CoP18) "Review of Significant Trade in Specimens of Appendix-II Species" (CITES 2002 [2019]), which outlines a process that has been developed to advise on whether trade would be detrimental to the survival of a species. Ninety days after the CoP, the Secretariat requests analysis of the CITES trade database (CITES 2002 [2019]). Species/country combinations are identified that warrant further scrutiny by the Animals and Plants Committees. The Secretariat notifies the range states of the species that they require further data about the trade and biological data of that species, and a preliminary report, including a categorization labelling the species as "less concern", "action needed", or "unknown status", is sent back to the Animals and Plants Committees (CITES 2002 [2019]). They determine if the categorizations are confirmed and make recommendations that are relayed back to the range states by the Secretariat. Compliance is measured by whether or not the range state has implemented the recommendations and within the timeline allocated by the Animals and Plants Committees. The Secretariat with the Animals and Plants Committees assesses the range state's compliance and shares this with the Standing Committee (CITES 2002 [2019]). Failure to complete or engage with this monitoring can result in a recommendation to suspend trade in that species from that country.

Review of Significant Trade is closely linked to Resolution Conf. 16.7 (Rev. CoP17) on "Non-detriment Findings" and Resolution Conf.14.7 (Rev. CoP15) on "Management of nationally established export quotas". Both of these elements are actions required by Article IV. NDFs are what takes place to ensure that trade is not detrimental to the survival of the species, as required in Article IV mentioned earlier (CITES 2013 [2016]). Conf. 16.7 recommends "Scientific Authorities should consider the volume of legal and illegal trade (known, inferred, projected, estimated) relative to the vulnerability of the species (intrinsic and extrinsic factors that increase the risk of extinction of the species)" (CITES 2013 [2016]: 2). Furthermore, the Scientific Authority is encouraged to choose an appropriate methodology for species in question, which may capture such data as:

a species biology and life-history characteristics;
b species range (historical and current);
c population structure, status and trends (in the harvested area, nationally and internationally);
d threats;
e historical and current species-specific levels and patterns of harvest and mortality (e.g. age, sex) from all sources combined;
f. management measures currently in place and proposed, including adaptive management strategies and consideration of levels of compliance;

g population monitoring; and
h. conservation status.

(CITES 2013 [2016]: 2)

The range of sources to collect these data could be literature, risk assessments, surveys, relevant knowledge and expertise of local and indigenous communities, consultations with the range of experts, and trade data from various sources (CITES 2013 [2016]).

Conf. 14.7 on quotas states that they are an important tool to conserve traded species, but since there is no Convention text that describes their use or implementation, the Resolution aims to establish a common understanding of how quotas should be managed (CITES 2007 [2010]). Quotas are annual limits on number or quantity of specimens that may be exported; they are not targets, so there is no need to reach the quota (CITES 2007 [2010]). An export quota system can replace the need for a case-by-case approach to NDFs. Quotas must take into account extenuating circumstances that may impact on species, like climatic conditions or other natural disasters, etc., which affect population numbers (CITES 2007 [2010]). NDFs should inform the setting of the national export quotas, and these quotas need to be well-communicated to the Secretariat, which will publish them on the CITES website (CITES 2007 [2010]).

The Review of Significant Trade appears to be an overarching monitoring mechanism that encompasses these individual elements to make sure they are contributing to the sustainability of the trade in Appendix II species. It is important to reiterate that all of these processes—NDFs, management of nationally established export quotas, and, technically, Review of Significant Trade—are guidelines as they stem from non-binding resolutions rather than the main Convention text. Clearly though, the Review of Significant Trade has become a cornerstone of CITES implementation as it now is a recognized element of compliance.

Belize, Benin, Cameroon, Cote d' Ivoire, Democratic Republic of Congo, Equatorial Guinea, Fiji, Ghana, Grenada, Guinea, Haiti (non-party), Lao People's Democratic Republic, Madagascar, Mali, Mozambique, Niger, Senegal, Solomon Islands, Togo, and United Republic of Tanzania are all subject to recommendations for trade suspensions of particular species because of a lack of a Review of Significant Trade. Thus, 19 parties (and one non-party, Haiti)—over 10 per cent—have trade suspended in particular species because the Review of Significant Trade has not taken place (CITES 2019). As an example from this list, since June 2010, Belize has faced a recommendation for trade suspension in *Myrmecophila tibicinis* (common name trumpet or Fluteplayer's Schomburgkia), an orchid species. According to the Notification to the Parties No. 2020/006, which provides all parties with the updated list of species- or country-specific trade suspensions for the year 2020:

> Parties are requested to inform their enforcement and customs authorities of the availability of the list of trade suspension recommendations on the CITES website in order to avoid the inadvertent acceptance of specimens of species subject to such a recommendation. Parties that require the issuance

> of import permits for trade in specimens of Appendix-II species are also encouraged to consult the list when processing applications.
>
> (CITES 2020: 1)

Thus, parties are largely expected to enforce trade suspensions against other parties that are *recommended*. Further research may be needed to assess how closely these recommendations are adhered to.

In cases of unresolved or persistent non-compliance, the Standing Committee sometimes suspends trade in all or certain species. In this regard, Article XIII—International Measures—applies. Application of Article XIII that is currently in effect is related to the Democratic Republic of Congo, Guinea, Lao People's Democratic Republic, Madagascar, and Nigeria (CITES no date). In the case of Nigeria, concerns have been raised about the legality and volume of trade to China, in particular, of *Pterocarpus erinaceus* (common name Barwood, or Rosewood; locally known as kosso), a threatened, endemic tree of West Africa (CITES 2018). The documents to the Standing Committee detail the history of the concerns and the actions and findings of the Secretariat's technical mission to Nigeria in 2018. Although there is much more detail in the documentation, in sum, the Secretariat:

> believes that a more effective "chain-of-custody" scheme to track timber should be established to ensure its legal origin and that the legality of the trade should be closely connected to the making of the Non-detriment Findings (NDFs). Currently, lax provincial regulations, loopholes in existing laws and the lack of sustainable forestry policies at the State level are exploited by national and foreign actors involved in the timber trade to export timber that is obtained in accordance with national laws but not in accordance with the Convention. In compliance jargon, this phenomenon is known as "legal optimization" or in other words "lawful but awful" trade conducted under the cover of genuine CITES permits.
>
> (CITES 2018: 4)

Whereas this may appear to indicate use of a significant "stick" to suspend trade, in actuality, the measures in place are to allow export after the confirmation of the authenticity by the Secretariat of the permits from Nigeria for kosso. A deeper dive into the trade data would highlight whether trade suspensions are in reality merely trade reductions, minor hurdles to business as usual, or actually complete suspensions.

Annual reporting and the Review of Significant Trade that stems from the trade data contained within those annual reports, as mentioned, are part of the compliance mechanisms of CITES. Another key element is that of biennial reports.

Biennial reporting

As one interviewee told me, there is an overall lack of implementation in regard to legislation and "shocking levels of compliance particularly in regard to reporting"

(CS3). Furthermore, this participant said that engagement with the resolutions and decisions "are just as bad". They thought this was, at least in part, because the Secretariat is hesitant to put timelines and deadlines in place, and this means that parties have no impetus to implement CITES (CS3). Whereas there are the supposed 90 days to put into place new measures and updated lists stemming from the CoP, in reality, there is no follow-up or way to force parties to act; there needs to be someway instead to incentivize implementation and compliance (CS3).

The lack of compliance may be most evident when analysing parties submitting their biennial reports. The failure to complete biennial reports is nothing new. Reeve noted back in 2002 that the biennial reporting requirement was largely unimplemented and that there was little follow-up. As noted earlier, Article VIII—Measures to Be Taken by Parties—sets out the elements of reporting (7a is quoted in the section on annual reporting):

> 7 Each Party shall prepare periodic reports on its implementation of the present Convention and shall transmit to the Secretariat:
>
> (b) a biennial report on legislative, regulatory and administrative measures taken to enforce the provisions of the present Convention.
>
> (CITES 1973: no page)

Also, as noted earlier, when discussing annual reporting, further guidance on reporting is given in Resolution Conf. 11.17 (Rev. CoP18) "National Reports". In the case of biennial reports, which are not specifically called that in the resolution, the Secretariat:

> 2 URGES all Parties to submit their reports required under the provisions of Article VIII, paragraph 7 (b), one year before each meeting of the Conference of the Parties, beginning with the 17th meeting of the Conference of the Parties and in accordance with the report format distributed by the Secretariat, as may be amended by the Secretariat from time to time with the concurrence of the Standing Committee.
>
> (CITES 2000 [2019]: 1)

Biennial reports have been required since 2005, when a report for 2003–2004 was expected. The CITES (2005) Notification No. 2005/035 sets out the purpose and the format of the biennial report.

> This format allows Parties to present information in a standard manner, so that it can be easily computerized, with three main objectives:
>
> i) To enable monitoring of the implementation and effectiveness of the Convention;
> ii) To facilitate the identification of major achievements, significant developments, or trends, gaps of problems and possible solutions; and

iii) Provide a basis for substantive and procedural decision-making by the Conference of the Parties and various subsidiary bodies.

(CITES 2005: 2)

The report includes five parts on general information, legislative and regulatory measures, compliance and enforcement measures, administrative measures, and general feedback. There are 105 questions, many of which are multipart. This may sound arduous, but many of the responses are check boxes, presumably to aid in the ease and quickness of completion.

After 2015, the format was changed, but the focus remains on the concrete steps parties have taken to implement CITES, their activities to ensure individual and organizational compliance, and their enforcement activities. These reports are now called implementation reports. The reports also assess the parties' perceptions of the difficulties in implementation of CITES. The biggest change is the clear link throughout the form to the Aichi Biodiversity Targets and the relevant indicators, which reframes the report as goal-orientated. Compliance with biennial reporting is as follows (beginning in 2003 through 2017):

Nearly 37 per cent of parties have *never* submitted a biennial report. Only 8 per cent (15 parties) have submitted all of them. From this information, it can be suggested that very little data about the implementation of CITES is being provided by the parties. Of the 15 parties which have submitted all seven of the possible biennial reports up to 2017, 14 of them have legislation rated as Category 1 by the NLP. My point here is that parties with legislation in Category 3 have no incentive to submit biennial reports. Since these reports break down implementation measures, parties which have not managed to implement the Convention may not want to highlight where they have fallen short by submitting their biennial reports. With this obstacle in mind, I asked in the Delphi iterative survey what could be done to improve parties' reporting.

As shown in Table 5.2, the participants of the first round of the Delphi iterative survey thought that a task force or working group to assist parties with reporting was the best of the actions suggested. Interestingly, the first choice about trade suspensions is essentially the existing system, yet 12 of 32 respondents disagreed

TABLE 5.1 Biennial report compliance

Number of Reports	*Number of Parties*
0	67
1	27
2	13
3	9
4	10
5	17
6	26
7	15

TABLE 5.2 What can be done to encourage member countries to complete their biennial reports?

Action	*Strongly Agree*	*Agree*	*Neither*	*Disagree*	*Strongly Disagree*
Introduce trade suspensions against those countries which do not report	9	8	3	10	2
Introduce reduced quotas against those countries which do not report	5	8	5	12	2
Create a task force or working group to assist countries with reporting	8	21	1	2	0
More visibility on CITES website and in the media about which countries do and do not report	8	17	3	4	0
Have a dedicated session at each Standing Committee meeting to complete reports	5	13	8	6	0
Have a dedicated session at each CoP meeting to complete reports	4	11	8	7	1

or strongly disagreed with this approach. This raises the need to further explore whether some parties are not compliant because they do not agree with the compliance mechanisms that are in place. I also suggested that a possible incentive could be to reduce the amount of trade of non-compliant parties by reducing or limiting the quotas, but this, like trade suspensions, received the least support (14 of 32 disagreed or strongly disagreed). After assisting parties with a working group, the action that was most supported was to have more visibility on the CITES website and in the media about whether parties are compliant with the reporting requirements. Twenty-five participants agreed or strongly agreed. Such "shaming" or highlighting good behaviour may motivate parties, which do not want any marks against their reputations. Drawing on the approach of other environmental conventions, in this case the Convention on the Protection and Use of Transboundary Watercourses and International Lakes (Water Convention) (A6), I proposed having dedicated time at either the Standing Committee meetings or at the CoP to complete the biennial reports. There was some support for this (18 and 15, respectively, agreed or strongly agreed), but there was also a high number (8–25 per cent) of respondents, who neither agreed nor disagreed. Perhaps a hybrid of having assistance or support to complete the implementation report at a scheduled meeting would help in compliance.

One interviewee noted, however, that "The CITES compliance mechanism already has a default suspension for countries failing to submit reports. While this generally encourages reporting, the quality of reporting is not always good. More focus needs to be on quality of reporting" (CS7). From the analysis of the biennial reports (and according to Bowman et al. 2010), this subjectively appears to be the case. Numerous reports have blank questions which have not been answered. For instance, in Chapter 4, when I talked about welfare, the question on the implementation report about confiscation of animals only had 35 of 69 participants providing an answer. Furthermore, some reports only have the check boxes filled in with no effort to provide commentary on the parties' actions or perceptions. In some instances, the biennial reports look exactly the same from multiple different time periods. It may be the case that nothing new has taken place in the 3 years between reports, but it could also be that no effort is being made to meaningfully engage with the process.

Participants of the first round of the Delphi iterative survey made suggestions as to other approaches that might improve the compliance in regard to reporting. One person proposed to "Exclude non-compliant parties form participation in some CITES activities until such time as they have completed reports". Another survey participant recommended: "Have a full report on non-compliance (there used to be one). Use of Article XIII by the Secretariat has been good, but too limited. Countries are too nice to each other". The latter is in a similar vein to another Delphi iterative survey participant, who thought "No excuse for not reporting. Makes a mockery of the whole system".

In the second round of the Delphi iterative survey, I followed up on some of these suggestions.

As is evident in Table 5.3, participants are not in favour of punitive measures. Thus, the suggestion to exclude non-compliant parties from CITES activities had 11 of 21 respondents who disagreed or strongly disagreed. Nearly all of the participants approved of the proposal for CITES to return to its previous practice of producing a full report on the non-compliance of the parties. Again, this might be an effective strategy, as it may have reputational implications if non-compliance is more widely known both within the CITES parties and in the wider world.

TABLE 5.3 What are your thoughts regarding the following?

Action	*Strongly Agree*	*Agree*	*Neither*	*Disagree*	*Strongly Disagree*
To improve compliance, exclude the non-complaint member from activities until they become compliant.	0	6	4	9	2
CITES should produce (as has been done in the past) a full report on non-compliance to share with all members.	11	9	1	0	0
The issue is not one of the quantity of reporting, but on the quality of reporting, which needs to be improved.	8	8	4	1	0

Related to a non-compliance report and overall reporting, in the second round of the Delphi iterative survey, one participant wrote:

> there should be more focus on compliance and less focus on enforcement under CITES; hence, many of the proposed options might be worth exploring within the limits of the convention. Parties with the capacity to fully implement and comply with the convention and the recommendations in the resolutions should do so to set the example. On reporting, the problem is more that so many reports and other information are asked of Parties taking away resources to do the work on the ground. A report on compliance may—or may not—be helpful. It depends how it is prepared and used by CITES.

This comment raises a number of points. First, the preparation and purpose of reports both from the parties and by the CITES Secretariat are worth considering. The move to electronic annual and implementation reports is partly intended to make these easier, but it is true that parties have numerous reporting obligations to CITES and possibly to dozens of other conventions and treaties. Second, this person mentions those parties which are able to fully implement and comply should do so to set a good example. For me, this links well with positive supportive efforts to improve the Convention. Furthermore, I suggest there is a connection to the suggestion from several participants of the survey and the interviews that importing countries should not import from countries that are out of compliance, that are not enforcing the Convention, or that have not properly completed Reviews of Significant Trade and NDFs. If parties with good practice were to play a more visible and significant role in holding other parties to account and expecting more from each other, this may be effective in improving compliance as well as taking some of the pressure off the Secretariat.

It is worth repeating that, as with implementation that I detailed in Chapter 4, in general, the participants of the survey and the interviewees supported approaches that tried to assist parties and help them to comply. For instance, one survey respondent said "Punitive measures are not appropriate for ensuring completion of biennial reports. Support could be provided to members who may have difficulty in completing the report through the secretariat or other appropriate body". Of course, to facilitate such support, CITES needs to receive more funding. The same can be seen in the next section, which explores CITES' own enforcement of its requirements.

Enforcement of the requirements

While the participants in this study supported non-punitive approaches to improving both implementation and compliance with the requirements of CITES, in general, they did seem to feel that CITES could do more to enforce its own requirements. For instance, one Round 1 Delphi iterative survey participant wrote: "More visibility on the CITES website in regard to non-implementation and non-compliance. Disciplinary action should be considered for consistently non-compliant members". As hopefully is evident after detailing ways in which implementation and compliance are not taking place, CITES is a highly complex multifaceted instrument. As such, I asked participants of the first round of the Delphi iterative survey, which component is the most important in regard to compliance? They were asked to score the level of importance of the components that I included from 1 (most important) to 5 (least important).

My reasoning for asking which component is the most important is to try to be able to make recommendations as to how to prioritize where improvements should be focused. With limited resources, there is only so much CITES and the Secretariat staff can do to try to increase parties' engagement with implementation and compliance. As one survey participant from Round 1 said: "Resolutions and Decisions are NOT binding on the Parties". And while I realize this, I included them in the components, as regardless of their non-binding nature, many of them form essential elements of effective implementation and compliance. With that in mind, CITES resolutions were scored as the second most important component (26 of 32 participants scored this as the most important or important) behind enforcing the legislation. This raises the question of what can be done by the Secretariat or the parties to better monitor Resolution implementation and incentivize parties to participate, given the nature of resolutions and their importance in the actual running of CITES. Is it possible that the most critical Resolutions should be incorporated into the text of the convention?

TABLE 5.4 Which component is the most important in regard to compliance?

Action	*Most Important 1*	*Important 2*	*Middle Importance 3*	*Somewhat Important 4*	*Least Important 5*
Annual reports being completed and submitted on time	6	10	8	5	3
Biennial reports being completed and submitted on time	4	9	7	8	4
Enforcing the legislation that is implemented	27	2	1	1	1
Enforcing the resolutions from the CoP	12	14	1	3	2
Enforcing the decisions from the CoP	10	9	4	4	5

Twenty-seven of 32 participants thought enforcement of the legislation was the most important compliance component. In regard to enforcing the Convention, this is an ongoing discussion. As I mentioned in the introduction, Oldfield (2003) has suggested that lack of enforcement may or may not be the issue, depending on what legislation is in place. Being able to enforce CITES means a party has properly prohibited violations and has a robust permit system in place. In Chapter 4, I shared that in my analysis of parties' legislation, 57 parties explicitly prohibit violations, 45 have some prohibition, and for 38, prohibition is implied by having penalties. There is a significant portion of parties who essentially may not have anything in their legislation to enforce. And enforcement is not explicitly required by the Convention. Having prohibition of violations as a requirement seems to imply enforcement, but as far as I can tell, there is nothing concrete directing parties that violations are to be looked for and that when they are found they have to be dealt with.

Interestingly, components that are required—annual and biennial reporting—were not given high scores in terms of importance. Perhaps this perceived lower importance of the two required reports on the part of the participants reflects the parties' own views. If the reports are not thought to be that important in the overall picture of what CITES requires, then it is perhaps not surprising that completion is poor.

Since Round 1 survey participants raised the issue of enforcement being so important, in the second round of the Delphi iterative survey, I asked: "One participant said that overall CITES does conserve species, but enforcement (weak or lax) stymies success. Do you agree with that statement?" Five respondents strongly agreed, seven agreed, one neither agreed nor disagreed, five disagreed, and two strongly disagreed. Of the 20 people who answered this question then, a majority of them thought lax or weak enforcement was keeping CITES from protecting species even better.

Both enforcement of the CITES requirements by the Secretariat and enforcement of the Convention by parties seem to be inadequate. The latter is beyond the scope of my study, but I touch upon it as it was clearly important to the participants. In regard to enforcement by CITES of the requirements of the Convention, a second-round survey respondent stated that part of the issue is the lack of robust and constant monitoring of parties by CITES. More resources are needed for the Secretariat to be able to undertake the level of scrutiny that is required for such a complex convention. That returns me to my point about prioritizing. While my participants scored enforcement of legislation as the most important compliance component, it is difficult to see how the Secretariat could monitor or enforce enforcement. Also, as I noted earlier, many parties do not have the proper legislation to enforce. Furthermore, the poor state of annual reporting means a lack of knowledge of the actual trends and scale of trade. Moreover, the poor implementation reporting means the Secretariat has limited information about the capacity and resources of the parties. Thus, I suggest the first steps to better enforcement are improving implementation (having the right legislation in place) and compliance (getting the right data from the required reports and reviews). It also seems that reporting and reviews are more achievable and less resource-intensive, and thus are able to be prioritized. More effort could be made by CITES to emphasize the importance of the annual and implementation reports as well as the Review of Significant Trade and NDFs. Resources put into improving these will have beneficial knock-on effects on enforcement. Then, trade sanctions can be reserved for the worst cases of non-compliance. But it is essential that non-compliance has some consequences. If there are none, then there is no reason for parties to comply. In Chapter 6, when discussing best practice and lessons learned, I return to what can be done to improve compliance.

Conclusion

Over one-third of parties have not reported on their implementation of the Convention since 2003. More than 10 per cent of parties are subject to some sort of trade suspension for failure to undertake a Review of Significant Trade. It is crucial then that the CITES Secretariat together with the parties take concrete steps to improve compliance. Supportive measures, such as working groups and direct assistance to complete annual and biennial reports, Reviews of Significant Trade and NDFs, should be combined with more active sanctions against persistently non-compliant parties. For instance, some of the trade suspensions have been in place since the early 2000s. One participant of the Delphi iterative survey Round 1 spoke to the interconnected nature of CITES and why each element informs the other and is therefore important: There is a need for "updated status of trade, model of efficient and effective implementation, level of changed threats and challenges should be shared and circulated amongst the compliance". The annual reporting provides the updated status of trade. The biennial (implementation) report provides models and examples of efficient and effective implementation as well as a space to share changing and emerging threats and challenges. Both of these reports are required, and it is time for parties to comply. And while the CITES Secretariat has the main role to play in improving compliance, parties themselves should take a more active and visible part in holding each other to account by requiring NDFs and robust reporting if trade is to take place. A transparent, fair, and strict approach to compliance by the Secretariat, supported by an equally transparent, fair, and strict approach by the parties, would help to improve compliance and therefore species protection.

References

Bowman, M., Davies, P. and Redgwell, C. 2010. *Lyster's International Wildlife Law*. Cambridge: Cambridge University Press.

Convention on the International Trade in Endangered Species of Wild Fauna and Flora (CITES). no date. CITES compliance procedures. Available at: www.cites.org/eng/prog/compliance. Accessed 3 September 2020.

———. 2020. *Notification to the Parties No. 2020/006*. Geneva: CITES.
———. 2019. Countries currently subject to a recommendation to suspend trade. Available at: https://cites.org/eng/resources/ref/suspend.php. Accessed 26 August 2020.
———. 2018. Application of article XIII in Nigeria. SC70 Doc. 27.3.5. Available at: https://cites.org/sites/default/files/eng/com/sc/70/E-SC70-27-03-05.pdf. Accessed 3 September 2020.
———. 2016. New implementation report format (formerly known as biennial report). Available at: https://cites.org/eng/node/43120. Accessed 11 November 2020.
———. 2013 [2016]. Resolution Conf. 16.7 (Rev. CoP17). Non-detriment findings. Available at: www.cites.org/sites/default/files/document/E-Res-16-07-R17_0.pdf. Accessed 3 September 2020.
———. 2007 [2010]. Resolution Conf.14.7 (Rev. CoP15) Management of nationally established export quotas. Available at: www.cites.org/sites/default/files/document/E-Res-14-07-R15_0.pdf. Accessed 3 September 2020.
———. 2005. Notification No. 2005/035. Available at: https://cites.org/eng/node/1759. Accessed 11 November 2020.
———. 2002 [2019]. Resolution Conf. 12.8(Rev. CoP18). Review of significant trade in specimens of appendix-II species. Available at: https://cites.org/sites/default/files/document/E-Res-12-08-R18.pdf. Accessed 3 September 2020.
———. 2000 [2019]. National reports. Available at: www.cites.org/sites/default/files/document/E-Res-11-17-R18.pdf. Accessed 26 August 2020.
———. 1973. Convention on international trade in endangered species of wild fauna and flora. Available at: www.cites.org/eng/disc/text.php#IX. Accessed 26 August 2020.
International Institute for Sustainable Development (IISD). 2019. Earth negotiations bulletin. August. Available at: http://enb.iisd.org/vol21/enb21101e.html?utm_medium=email&utm_campaign=ENB%20Update%20%2030%20August%202019%20SW&utm_content=ENB%20Update%20%2030%20August%202019%20SW+CID_a7fade43500fa9f1ca5d10a7a4d2b3fa&utm_source=cm&utm_term=Read. Accessed 21 October 2019.
Oldfield, S. 2003. Introduction. In Oldfield, S. (ed.). *Trade in Wildlife: Regulation for Conservation*. London: Earthscan: xvii–xxvii.
Reeve, R. 2002. *Policing International Trade in Endangered Species: The CITES Treaty and Compliance*. London: Earthscan.
Wijnstekers, W. 2018. *The Evolution of CITES: A Reference to the Convention on International Trade in Endangered Species of Wild Fauna and Flora, 11th Edition*. International Council for Game and Wildlife Conservation. Available at: www.cic-wildlife.org/wp-content/uploads/2018/05/the-evolution-of-cites-2018-11th-editionpart-1.pdf. Accessed 27 November 2019.

6

LESSONS LEARNED AND BEST PRACTICE FROM THE CASE STUDIES

The previous two chapters have detailed my project's findings in relation to implementation of (Chapter 4) and compliance with (Chapter 5) CITES. At times, this can paint a bit of a bleak picture as there is substantial room for improvement. That said, the main objective of my research on CITES was to identify best practice of and lessons learned by the parties, so that other parties can see concrete paths to achieve such improvement. Thus, this chapter focuses specifically on sharing the best practice and lessons learned that were identified through the Delphi iterative survey and the 20 expert interviews. I not only draw on the case studies that were chosen from the Delphi iterative survey, but also detail my expert participants' thoughts on best practice for implementation of and compliance with CITES in general. I also provide further insights from the legislative content analysis.

To begin the chapter, I relay in more detail than in the methodology chapter how the case studies were chosen. I then breakdown my analysis of best practice and lessons learned into sections corresponding to the previous chapters. Thus, I discuss the four components of the NLP—authorities, prohibition, penalization, and confiscation—in one section before talking about overall best practice and lessons learned in regard to implementation. Then, I turn to an analysis of compliance measures that were mentioned by experts as being best practice. Finally, I discuss the best practice and lessons learned, specifically taken from the three case studies—Canada, Indonesia, and South Africa—before providing some further thoughts on how to improve CITES. These further thoughts on improving CITES revolve around the science that is supposed to underpin the Convention, holding parties more accountable in regard to implementation and compliance, making sure all the relevant stakeholders are involved, and modernizing the Convention.

Case studies

In Round 1 of the Delphi iterative survey, I asked in Question 23 "Which country has a good approach to implementation of and compliance with CITES"? In Question 24, I asked "From which country could other members learn lessons"? Only eight and six participants, respectively, out of a total of 32 participants answered these questions. A vast majority of the participants listed the European Union and the United States for both questions. This is not unexpected. As one participant pointed out, wealthy Western parties tend to have strong implementation and compliance because they have the resources (both financial and human) to devote to them. While a demonstration of what the commitment of resources can lead to in terms of implementation and compliance may be useful, I do not think it is replicable for a vast majority of CITES parties. Thus, even though the EU and US were singled out as good examples, I decided that their approach to implementation and compliance (and enforcement) would not make the best case studies for how other parties might improve their own implementation and compliance.

Instead, I chose to focus the three case studies of my research on three CITES parties that have received limited attention, but which were identified by participants in the first round of my Delphi iterative survey as doing some element of CITES well or having improved their implementation and/or compliance. This led me to having Canada, Indonesia, and South Africa as the three case studies. Many of the interviews that I conducted focused on these three parties in particular, so that I could better understand what the parties were doing well and how their approaches might provide guidance for other parties to improve. The interviews did, however, also focus on CITES overall, which, as I mentioned, I conclude the chapter with. As mentioned, this first chapter section contains further insights on the four components of the NLP, so first—authorities.

The NLP

The first component—authorities—is where my expert participants seemed to have the most to say. There was no consensus on whether there was a particular agency or section of the government that was most suited to housing the authorities, or even whether or not the Scientific and Management authorities should be separated. But participants did agree that the authorities need to be independent. "There must be strong independent authorities that do not influence each other. For instance, a Scientific Authority that is manipulated by the Management Authority means quotas are meaningless" (Delphi iterative survey Round 1 participant).

One interviewee noted best practice for the authorities is when CITES implementation is viewed as a partnership (G4). "The authorities work closely, but with their separate remits. They [the SA and MA] have clear roles and know when to include the EA [Enforcement Authority]" (G4). In this particular case (Canada), new staff are cross-trained by all three authorities. Furthermore, the authorities have access to each other's systems where possible. For instance, the Enforcement Authority can, in certain instances, see permits issued by the Management Authority, and the Management Authority can assist with intelligence analysis (G4). A partnership approach takes on a different form, but important role in South Africa. For countries (like South Africa among others) with provinces or states and decentralized approaches to CITES, a committee approach to the Scientific Authority allows for representation of all the provinces (CS8). In the case of South Africa, the committee also has a member of the Management Authority to ensure collaboration (CS8).

Several participants thought best practice was for the authorities to be specialists with experience and expertise with wildlife/CITES (A2, G4, G5). According to one interviewee, you need to deal with CITES full time to grasp the subtleties (G4). Furthermore, in many parties there are also personnel changes so there is no continuity to implementation (or compliance) (G3), which is another element to address in terms of best practice. I mentioned this in Chapter 3 when contacting all the authorities to take the Delphi iterative survey, it was clear, from the number of broken contact links, that there is a high turnover rate in CITES staff around the world. Regarding the authorities that are actually needed to implement CITES, as mentioned in Chapter 4, in general, participants felt adding an Enforcement Authority would be useful. In contrast, one participant of Round 1 of the Delphi iterative survey said, "The Scientific Authority is the most critical, but the focus in on an enforcement authority". An interviewee noted:

> In some cases—A voice that is missing from Government is the voice of law enforcement officials—a key recommendation would be to ensure the decisions taken by the MA are actually enforceable and engage with the agencies [law enforcement] tasked with doing this.
>
> (CS6)

Thus, reiterating the key elements of best practice are independence and partnership of the authorities.

In regard to prohibition, one Delphi Round 1 participant identified the EU as engaging in best practice in regard to prohibition by implementing stricter rules than required by CITES. By treating all species like Appendix I, thus requiring import and export permits, trade and trafficking are thought to be better monitored. Similarly, one interviewee observed that while India, for instance, does not implement all four of the NLP's components, they do well combatting illegal wildlife trade. This is because of domestic restrictions which protect Indian wildlife (CS1). Thus, prohibition at the national level, for some experts, is linked to best practice in terms of CITES implementation. As quoted in Chapter 4, for some, however, this goes against the spirit of the Convention.

Similarly, for penalization of violations of the Convention, there was no consensus as to what approach to penalization constitutes best practice. Numerous participants noted that because of the varied jurisdictions, judicial systems, and cultures, a universal approach is impossible. One suggestion, however, that I argue has traction across all parties, is that a portion of fines or proceeds from asset seizure should go back into conservation and wildlife protection measures. The legislative content analysis revealed that numerous parties have an environment or conservation fund. These could be used as a repository for fines, but also for alternative sanctions like asset forfeiture and/or seizures. Further research would be needed, however, as to the best way to structure such funds and their effectiveness. Fines are the most common form of punishment (see Chapter 4), and the power for asset forfeiture and/or seizures for wildlife crimes could symbolize the importance of them to law enforcement agencies. These sources of income would be available to all parties and could provide the needed additional resources for implementation and compliance at the national level. The caveat is, of course, that sufficient measures need to be put into place to protect the capture of these funds by corrupt actors.

Finally, in terms of the NLP components, there were a number of recommendations for what best practice in relation to confiscation entails. These link directly to the penalties' best practice, as the fines and proceeds from seized assets could be used to pay for these approaches. First, in terms of wildlife products, there need to be proper evidence procedures in place as well as appropriate storage facilities. Several participants in the Delphi iterative survey commented that there must be regulations in place that these products (or live wildlife) cannot be resold. I suggest the reasons behind this are two-fold. One, not selling the wildlife or wildlife products challenges the norm that wildlife are commodities. Thus, perhaps, there is a symbolic element to not selling the wildlife and wildlife products. Two, not selling the wildlife and wildlife products removes the temptation corrupt law enforcement may have to erroneously confiscate the wildlife and/or profit from the sale. Second, in terms of wildlife who are alive, CITES parties must cooperate to return the wildlife to the origin country wherever possible, even though, as noted in Chapter 4 during the reporting period from 2015 to 2017, this was only accomplished six times according to the parties' own accounts. Where the wildlife cannot be repatriated, parties need to have memorandums of understanding in place with reputable partner organizations (zoos, rescue centres, botanic gardens, and so forth) to transfer the wildlife to these facilities. Again, the fines and seized assets could cover the costs of the storage facilities, transportation, and/or housing and care. Some parties shared that cost was the challenge in regard to confiscation.

Overall, the four criteria also do not fit across the parties (CS1). Small island nations, for instance, have a different experience and ability to implement and comply with CITES than parties like India or Viet Nam, which are source, transit, and destination places for large amounts of wildlife. Furthermore, across the overall assessment, those undertaking that assessment to categorize CITES parties are very generous (CS7). My participants felt, though, there is no interest in changing or improving the NLP or to give greater scrutiny to parties listed in Category 3. Also of concern is that once a party's legislation is graded as a Category 1 party, "there is no incentive to improve even though the legislation could be better" (CS7). Thus, even Category 1 parties may have problematic implementation, as seen in Chapter 4, when briefly discussing Madagascar. Or, as is also the case in Madagascar, the party may be Category 1 and be non-compliant, so even "good" implementation may not lead to effective or successful working of the Convention. As with so many decisions and policies, the NLP is making progress, but it is too slow (CS7). And, similar to the CITES parties themselves, as seen in Chapter 2, there is no reliable funding mechanism for CITES overall, with the contributions by parties being non-mandatory and the amendment for the budget not being ratified by all the parties. This is in contrast to the Global Environment Fund and the Convention on Biological Diversity, which have more secure sources of income and, therefore, are more likely to be implemented.

As one interviewee said to me, "If there was no NLP, it would be worse" (CS7). While the NLP does provide some degree of oversight and lead to some good practice in terms of implementation, there remain areas for it to improve, as I touched upon in Chapter 4. In particular, "The NLP is flawed, though, as it doesn't adequately deal with the enforcement angle, and it really only judges the legislation against terrestrial species rather than also ensuring legislation for marine species meets the four criteria" (CS7), as highlighted in the next section. The latter portion of this statement about only judging the legislation against terrestrial species raises an important point about the legislation itself. It seems to me that the NLP starts from a flawed assumption that parties have legislation that incorporates CITES' definition of wildlife. To evidence what I think is a fundamental problem with CITES parties' legislation, I return to my legislative content analysis and unpack the very definitions of wildlife underpinning the range of parties' legislation.

Legal definitions of wildlife

As part of the legislative content analysis, with some help from Alison Hutchinson, I recorded how each piece of legislation defined wildlife. In Table 6.1, all the variations of definitions of wildlife can be seen. Again, there is the caveat that there may be some inaccuracies due to translations, but I suggest these translation differences are unlikely to greatly affect the exclusion or inclusion of fish and plants in the definitions of wildlife, which are the areas that are particularly striking. This content analysis was completed nearly a year after the original analysis. During that time, five links to CITES' legislation of parties ceased working, so there are nine pieces of legislation where we were unable to determine what the definition of wildlife is.

In 32 countries' legislation, fish are specifically included in definitions of wildlife or animal. In 14 instances, however, they are expressly excluded. "Partially included" in Table 6.2 means that only certain fish were included; in these cases, there was a distinction made that river species were included, but no mention was made of saltwater species. The seven cases where we determined fish are probably not included is in legislation where animals or wild animals are listed (i.e. mammals,

TABLE 6.1 Overall definitions of wildlife

Definition Code	*Count of Definition*
Animal	17
Animal and specimen	1
Animals and birds	1
Animals and plants	31
Animals, plants, birds	1
Biodiversity	2
Biological diversity	1
CITES	10
Fauna and Flora	31
Fauna	3
Living resources (animals and plants)	1
Natural resources	2
None	33
Specimen	1
Wild animals	2
Wild species (animal and plant)	1
Wildlife	26
Wildlife and biodiversity	1
Wildlife and fish	2
Wildlife and forestry	1
Wildlife and plants	1
Wildlife and timber	1
Wildlife, forestry, and fisheries	3
Wildlife, genetic resources, forests, fisheries	1
Not able to access text	9
Grand total	183

TABLE 6.2 Breakdown of definitions of wildlife analysing fish and plants

Wildlife	*Codes*							
	Included	*Excluded*	*Partially Included*	*Prob Excluded*	*Implied Included*	*Timber*	*Separate*	*No Definition*
Plants	69	10	3		5	2	15	81
Fish	32	14	10	7	61		8	53

birds, and reptiles), and no mention of fish is made. The code "Implied Included" are definitions that say all or any animals and discuss regulation of fishing, fisheries, and/or law of the sea, which we interrupted to mean fish are included in wildlife. The "Separate" code indicates that the "wildlife" legislation does not necessarily include fish, but there is another piece of legislation regulating fish and/or fisheries.

Plants appear to be included in wildlife in 69 pieces of legislation. They, too, are specifically excluded in 10 cases (see Table 6.2). As with fish, there are examples of plants partly being included in definitions of wildlife. In this legislation, inclusion was of native plants and/or plants growing naturally (rather than being propagated). Plants are implied to be included, when legislation refers to forestry and harvesting. In only two cases, timber was specifically listed as being wildlife. For eight pieces of legislation, there was obvious reference to separate (probably forestry) legislation. In total, then, there are 38 cases where neither plants nor fish are mentioned in definitions, and, in 17 cases, both plants and fish are specifically included.

I suggest the exclusion or inclusion of certain species and not others in legislation has significant implications for the implementation of CITES. As Alison, another colleague, and I argue elsewhere (Wyatt et al. Under review), fish often seem to be absent from debates about wildlife management and wildlife trade. This has potentially profound consequences, considering fish are not only the largest portion of non-human animal life, making up almost three-quarters of the weight of all non-human animals (Bar-on et al. 2018), but also that fish are some of the most traded food items in the world today, with USD 130 billion in global annual trade (UN Food and Agricultural Organisation 2020). For fish, then, to not be defined by some parties as wildlife could mean that fish would not be afforded the same protection as other species, were they to be listed on the CITES appendices. It seems to be clear that the inclusion of fish or not, though, is not taken into consideration when gauging whether a party has fully implemented the legislation for CITES, as the participants of this study pointed out to me and which I talk more about in Chapter 7.

The same problem exists for plants. If they are not protected to the degree necessary to prevent their extinction because they are not listed at all or because they are not included in "wildlife", this has troubling consequences for ecosystems and the biodiversity crisis. Margulies et al. (2019) call out this "plant blindness" as impacting upon the policies and research studies that would effectively protect more plants than just timber. This may seem surprising, considering a vast majority of CITES-listed species are plants (over 30,000—as mentioned in Chapter 2), but becomes more apparent when it is made clear that nearly all of the plants listed are orchids.

Thus, in addition to the improvements that could be made around the four components to the NLP, there is work to be done in the CITES legislation even a step before the assessment of the implementation. To truly stem the biodiversity crisis, what wildlife is needs to be clear and enshrined in the legislation that is providing protection. This is essential to both sustainable use and to protection. I turn now to the data I gathered about what best practice in the overall implementation of CITES legislation looks like, not just the NLP components.

Overall best practice for implementation

Not part of the NLP, but clearly critical to the functioning of CITES, is the permit system. Participants offered numerous suggestions as to what constitutes best practice for permits as well as overall approaches to the implementation of CITES.

In terms of permits, it was suggested that there needs to be set criteria for permit issuance that are strictly adhered to regardless of whether permits are centralized or devolved to provinces or states (G4). However, a centralized system is better to ensure consistency. In designing a system to use CITES permits, the quality assurance process should be built in (G4). One suggestion was that each permit has to be seen by more than one person to create checks and balances (G4).

A few interviewees recommended that permits were an obvious way of funding CITES implementation and compliance at the national level. "CITES has complex reporting and is expensive, but why not have members charge more for permits and/or tax trade—trade isn't necessary, so if you can't cope use permits as the funding to support proper implementation and compliance" (CS7). Another interviewee noted: "[There are] poor people but not poor countries, particularly not for traders and forestry officials. Charge exporters to get licenses and then fund complete proper NDFs. Have a general fund then independently do the work [of the Non-detriment Findings]" (A3).

There was a general consensus that best practice for permits has evolved to require some form of electronic permits (CS6, G2, G4, numerous Delphi iterative survey respondents). There are several reasons why this is best practice. First, without an electronic permitting system, there are risks of human error on data capture (CS6). Thus, e-permits are thought to improve the quality and consistency of reporting, which is an ongoing issue for CITES (Wyatt 2013). Second, paper permits also make it difficult to ensure effective transparency and the application of checks and balances (CS6), which are part of best practice, as outlined earlier. One interviewee suggested looking to Switzerland as a model for e-permits as their luxury goods trade (e.g. watches) uses an electronic system that tracks every move of each item (G3). As reiterated several times, CITES is a complex convention to implement and comply with. One of the several factors that contributes to this complexity is that validation of a commodity (i.e. wildlife) at export is contrary to how all other trades work; this creates confusion and hurdles (G4). An e-permit system may also help to reduce the complexity by simplifying the procedure, which would also help to reduce confusion.

Similarly, one interviewee stressed the importance of undertaking robust NDFs (G4). Again, these should have set criteria to ensure they all are completed to a high standard with all of the required information and input. Since wildlife trade and CITES are international, one participant suggested that parties should try to coordinate an inter-regional approach to CITES implementation. This would entail neighbouring countries harmonizing their CITES legislation (G5). This would address a variety of issues such as wildlife traffickers targeting the locations with the weakest legislation and law enforcement and government authorities struggling to work together across different systems.

A unique suggestion for best practice was for CITES authorities to be empathetic to the people asking for permits. It was recommended that "You [CITES Authority] need to understand the lived realities of the wildlife consumers and industries. Go and talk to people—don't be bureaucratic" (G4). For a complex and highly bureaucratic convention, this is a difficult, but poignant, piece of advice. It also touches upon the intractable tension within CITES that I mentioned previously between people advocating for use of wildlife and those arguing for protection. The suggestion and the tension—talking to wildlife consumers and industries and utilization versus protection—coincide in many ways, but in particular when considering indigenous people's use of wildlife.

For example, as related to me by one interviewee (G5): Canada has been identified as improving the CITES authorities' relationship with First Nation communities. This was achieved by incorporating indigenous communities into the governmental and CITES decision-making. The need for improvement was sparked by controversial listing proposals in the 1990s about the polar bear. These controversial proposals forced Canada to include indigenous members on co-management boards that include members of the federal government, the indigenous communities, and the provincial governments. One focus of the boards is on how commercial export of Canadian protected species can be done sustainably, fairly, and transparently. The export of commercial species, like the polar bear, has a three-pronged approach around traceability. When a polar bear is killed, the person who killed it contacts the conservation officer. The conservation officer then microchips the various parts of the polar bear that are destined for export. The conservation officer takes DNA samples and stable isotope samples to pass along to the scientists. This ensures traceability from the kill to the sale and has been designed with the input of First Nation hunters, who rely on the proceeds.

The Canadian approach is just one example of the importance of the human relationships that underride all the bureaucracy of authorities, prohibition, penalization, and confiscations. And the example highlights that best practice for implementation is not just about technical fixes and legislative wording, but people working together to ensure fairness for human communities and consideration of other species. This is applicable to compliance with CITES as well, but there are other elements to best practice in terms of compliance too.

Compliance

The NLP checks on the implementation of the core legislative components of the CITES convention. Failure to implement then becomes a compliance issue, and this section explores what can be done to improve compliance with the implementation as well as compliance with the reporting requirements detailed in Chapter 5. As mentioned repeatedly, all the various elements of CITES suffer from lack of funding and the NLP is no different. As one interviewee told me, "the NLP is not well funded, so its work has been bity" (G3). Regardless of the funding, there is the need for a uniform application and transparency in terms of the compliance measures that arise from the NLP. Furthermore, compliance measures should not only be targeted at parties with the most glaring problems; even parties with low trade levels, which are failing to implement CITES legislation and comply with its requirements, should be considered for compliance measures. But, as detailed in Chapters 4 and 5, it is not always clear how the NLP categorizations have been arrived at and whether or not they are consistent, and the same is true of when and whether or not sanctions are applied.

Furthermore, lessons learned in regard to these compliance measures are that "There has been legislation progress, but the sanctions for not implementing or complying don't bite" (G3). This is particularly true when the CITES party has low volumes of trade and/or they have high levels of illegal trade (G3). Illegal trade circumvents CITES already and, in that case, and instances of low trade volumes, suspensions matter very little (G3). Thus, these factors need to be taken into account when trying to ensure compliance of certain parties. Other factors that need to be taken into consideration are that some countries have no capacity and/or no incentive to change (G3). Again, this may require an approach different from the previous efforts to achieve compliance.

Of particular concern, and certainly requiring a different approach from CITES, is the claim by some of my interviewees that some parties have not gotten more compliant but are instead smarter at hiding their non-compliance (A3, CS5, CS10). As one interviewee said to me:

> They [parties to CITES] have learned how to tell the right falsehoods and lie to make it look like they are implementing CITES as it should be. They have also learned not to let outsiders in or NGOs, so it is easier to say there are no problems. People [CITES and other stakeholders] are willing to believe the nonsense . . . The EU should be asking for better proof, evidence. It is very complicated though when the MA of a country claims that all has been checked. And sometimes this is backed up by the SA. CITES is difficult—blatant lying is nearly impossible to combat. It is good for business to believe they [parties] do check. The EU does well, but is constrained because CITES is a trust issue. They ask questions, but if they [the exporting party] say yes, how can you not believe them?"
>
> (A3).

The "check" referred to by this interviewee is when the importing party asks to see the NDF (discussed in Chapter 5) of the exporting party. A seemingly insurmountable problem arises if the exporting party has (purposely) completed the NDF poorly or falsely to ensure that trade looks sustainable. Thus, as mentioned earlier, quotas that are linked to NDFs must be set independently with no input from exporters (A3). And once quotas are set, they need to be followed and monitored (A3). NDFs highlight, once again, the need for the authorities to work together, but independently. They also raise the issue of the need to consult with the range of stakeholders beyond only the authorities. As one interviewee noted, participation must mean actual consideration of and listening to participants:

> In accordance with the provisions of Regulation 4(3)(b) of CITES regulations, the Scientific Authority must consult, when necessary, with the private

> sector, non-governmental organizations, local communities, and other stakeholders before making any findings or recommendations or giving advice. (CS6)

CITES, then, has best practice built into it by mandating collaborative, consultative processes, but there are numerous examples where parties fail to adhere to these regulations. The final point regarding implementation best practice being about the human relationships and working together also applies here about the best way to approach compliance. This is particularly the case, when, as described earlier, trust is central to the interactions.

Recommendations on improving compliance

- More support for completing annual and biennial (implementation) reports either through a mentor system or support sessions at Conferences of the Parties or committee meetings.
- Stricter enforcement and adherence to the deadlines by the Secretariat and the Standing Committee for reporting requirements (3 years is a significant, and possibly risky for the wildlife, delay).
- More visibility as to compliance and non-compliance with CITES meetings (non-compliance report for the parties) and on the CITES website.
- Stricter enforcement by the Secretariat and the Standing Committee of Reviews of Significant Trade and greater scrutiny and standards from other parties for accepting them and the NDFs.

More specific examples of best practice and lessons learned are found in the next section detailing the case studies—Canada, Indonesia, and South Africa.

The case studies of lessons learned and best practice

Canada, Indonesia, and South Africa all have Category 1 legislation according to the NLP. Thus, there is some similarity between them, but, in general, these parties' approaches to implementation and compliance provide different ways in which best practice can be achieved. And even though all of these parties have Category 1 legislation, which indicates they have fully implemented CITES, this in no way means that any of the three have perfect legislation and that there are no areas for improvement. This in itself supports the previous mention that the NLP, while useful, is limited in its ability to push for continued improvement in regard to implementation of CITES.

Case study 1—Canada[1]

As indicated, Canada was listed by Delphi iterative survey participants as a party that has implemented CITES well overall. There were, however, areas for improvement from which other parties could learn. Two interviewees (G4, G5) noted that wildlife was not a priority of the border agency in Canada and that this hampers enforcement. This was mentioned in terms of implementation in the context that the CITES authorities in Canada (the Canadian Wildlife Service, which includes the Management and Scientific Authorities and the Wildlife Enforcement Directorate, which is the Enforcement Authority) do not always work as closely with the border agency as they do with each other. This is certainly not unique to Canada. Several interviewees and Delphi iterative survey participants noted the challenge of overcoming the lack of priority of border and customs agencies around the world in regard to wildlife.

The main lesson to be learned from Canada stems from its cumbersome approach to recognizing species who are newly listed on the CITES appendices after a CoP. Canada's legislation must be updated each time a new species needs to be listed. Updating legislation requires public consultation and several readings of the legislation in the legislature. Thus, there is essentially no way that Canada is able to add new species to their legislation within the 90 days after the close of the CoP, which is required by the Convention. Hence, after each Conference, Canada enters Reservations against the newly listed species until such time as the legislation is updated to reflect their protected status. The lesson from this is that the legislation implementing CITES is most efficient and effective if written in such a way that the CITES lists in total are recognized and thus updated automatically, rather than the national lists having to also be updated each time a change is made.

Apart from the low priority of wildlife for the border agency and the approach to recognizing the CITES-listed species, Canada has several elements of good practice in terms of implementation and compliance, which could help other parties to improve. In particular, as mentioned earlier, Canada has three authorities. Each of these is distinct with clearly defined roles and only work on wildlife. In Canada, this means CITES as well as national and provincial wildlife legislation, but the specialization on wildlife was touted as a strength of Canada's approach to implementation. Furthermore, these three authorities have a close working relationship (G4, G5). One interviewee thought this close working relationship was fostered by the fact that the three authorities are located within the same office building (G5). Presumably, the physical closeness helps to build the personal relationships that form between the staff of the authorities. This may indeed contribute to their good working relationship and, thus, their effectiveness.

The importance of good working relationships extends beyond the three authorities. My interviewees noted that staff of the Canadian Wildlife Service and the Wildlife Enforcement Directorate as well as Environment and Climate Change Canada in general (the government department under which all three authorities sit) also have good working relationships with other Canadian government departments and convention authorities, such as the Convention on Biological Diversity. Furthermore, the Canadian CITES team has good working relationships with international partners, particularly the United States Fish and Wildlife Service, which is the CITES team in the US. This is particularly important in the context of enforcing CITES, as the US is one of the largest consumers and re-exporters of CITES-listed wildlife and the main trading partner with Canada. In addition, the Canadian CITES team is very active on the international stage. They have held influential positions at INTERPOL, for example, serving on the Wildlife Crime Working Group. They also actively engage with proposals and species listings at the CITES CoP. Such engagement leads to in-depth knowledge of the Convention and presumably helps in its national implementation.

In Chapter 4, I mentioned that some Delphi iterative survey participants and some interviewees suggested that stricter national measures for protecting wildlife resulted in better implementation of CITES (and protection of species). Canada is an example of this, as they have an added layer of scrutiny at the provincial level. The legality of the harvesting, capturing, and/or hunting of wildlife is checked for movement within Canada, and permits are issued at this provincial level. One disclaimer to this seemingly comprehensive approach is that Canada has very few native species listed in the CITES appendices. As one interviewee pointed out, it is not an arduous administrative task for Canada to be able to do this, when for other parties, this would likely be impossible (CS7). That is not to say, however, that this approach does not provide an example for some CITES parties.

A further element of best practice seen in Canada's approach to implementation is their incorporation of indigenous communities in decision-making, as outlined earlier. The inclusion of the First Nations people on committees, and so forth, is worth further investigation to better understand how this can be improved upon even more. Finally, as with probably all parties, Canada has limited funding dedicated to the implementation of CITES and its enforcement. Interviewees suggested Canada compensates for this limitation by engaging in multilateral operations. Thus, by partnering with other CITES teams in other countries, expenses can be shared. This, too, is worth further exploration to see if there are specific areas of implementation and compliance that would benefit from a collaborative approach between parties (e.g. e-permits, reporting, among other possibilities).

Summary

Other parties to CITES can potentially learn a few things about implementation from Canada. Of particular importance is the need for a legislative scheme that updates automatically to reflect listings quickly and easily. Canada's experience also reiterates the continued need to include border and customs agencies in the collaborative efforts of the CITES authorities. Canada's approach to CITES also presents three possibilities for other parties to consider. First, there appears to be strength in having specialized CITES staff or at least people specializing in wildlife in general with CITES as part of their remit. Second, there are benefits to establishing good working relationships between the authorities, with the local and indigenous communities, and with other countries. Third, there is value in actively getting involved in global activities, be it CITES committees or conferences, INTERPOL working groups, or cross-border operations. There is some overlap, but also a different set of lessons, and best practice can be taken from Indonesia's approach to CITES implementation and compliance.

Case study 2—Indonesia

The overlap of elements between Canada and Indonesia is the scrutiny of intranational or domestic movement of protected species. In the case of Indonesia, this is a much more difficult administrative task since the number of native CITES species protected in Indonesia is much greater than in Canada. What the two do have in common is the scale of territory that needs to be monitored. Both countries are vast in terms of square kilometres, though obviously they have very different topography and ecosystems. Indonesia, having tens of thousands of islands and marine ecosystems, presents different challenges to the tundra and arctic landscapes of Canada. In both instances though, oversight of the movement of nationally protected and CITES-listed species was said to improve the implementation of CITES. Perhaps this is due to the fact that tracking internal movement adds a layer of evaluation or point in the supply chain to ensure that the proper procedures are in place.

As mentioned here, Indonesia has significant levels of CITES trade. Each year, there are more than 5,000 records of exports and more than 500 records of imports (Wyatt 2020b). These are only *records*, so they do not indicate the actual volume of trade or number of individual wildlife, which are both much higher. With such high volumes, it is important that there is a robust and rigorous setting of annual quotas (G2). The process of setting quotas raises, again, the importance of NDFs, which are directly linked to quotas. Two particularly important elements come up here, both of which have been touched upon in other parts of this book. First, those with a financial interest in the quota that is set (higher quotas mean more permits are issued and more wildlife can be sold, thus increasing profits) should not be able to influence the setting of the quotas. Interviewees expressed concern that traders and exporters, in Indonesia specifically, hold too much sway in the quota setting process (A3, CS5, CS7). Second, if there is any doubt on the part of the importing party that the export of CITES wildlife threatens that species survival then the importing party should push back and ask for evidence of NDFs (A3, CS7). Interviewees indicated that Indonesia has made noticeable improvements in relation to quotas and NDFs (CS7, G3), but there are lessons to be learned for other parties in terms of how quotas and NDFs can be made even better.

A further lesson to be learned, but perhaps more so a warning of an aspect of trade that is in need of urgent attention, is the apparent lack of oversight of captive breeding (and artificial propagation) that is on display in, but not exclusive to, Indonesia. For wildlife traders, captive breeding and artificial propagation are often thought to be sustainable money-making alternatives to procuring individuals from the wild. There are some examples where this is the case, such as the captive breeding of crocodiles for leather in Mexico. So, there is evidence and promise that such establishments might provide the needed income to people reliant on wildlife economies. What can be learned in the case of Indonesia, however, is the critical need for such businesses to be properly monitored and evaluated.

This has been most thoroughly documented by Nijman and Shepherd (2015). In their assessment of commercial breeding of Tokay geckos, a number of factors are present that can be used to develop monitoring guidelines. These include the scale of breeding being undertaken. In the Tokay gecko facilities, the business operators were given permission to breed up to 3 million geckos.

> In order to produce one million adult-sized geckos a facility would require 140 000 breeding females, 14 000 breeding males, 30 000 incubation containers in continuous use year-round, and some 112 000 rearing cages. Basic care of these Tokay Geckos would require hundreds of staff to be employed and a constant supply of food, all of which would have significant additional costs.
>
> (Nijman and Shepherd 2015: 1)

The fact that no such facilities existed was a clear indication that such breeding was not taking place and, in reality, the geckos were being taken from the wild. Furthermore, there was no previous experience or history of breeding operations on this scale, another indication that breeding was not likely to be taking place.

Other factors that could be included in monitoring such facilities are the reported purpose of the trade and the type of business that is seeking to breed wildlife. The 3 million Tokay geckos were supposedly destined for the live pet trade, but there was little evidence that there was that much demand for geckos as pets. But there was data to support that much demand for dried geckos as an ingredient in traditional medicines. In terms of the type of business, the companies seeking permits to breed geckos were not pet traders, but suppliers of traditional medicines. The purpose of trade and the type of business link to Nijman's and Shepherd's (2015) final point around expenditures and profits. The profits from the pet trade, even of 3 million geckos, were not estimated to cover the costs of running a commercial breeding facility of the scale required. Thus, the facilities were highly likely to be laundering wild-caught geckos.

Summary

Admittedly, this is not a flattering picture of CITES implementation. There are, however, numerous lessons to be learned. The pitfalls and loopholes of captive breeding and artificial propagation become apparent, but monitoring the scale, purpose, expenditures, and profits of the operations, and the experience and history of the operators may help to prevent laundering and contribute to protecting wildlife. In addition, the importance of transparent and independent quota setting as well as the importance of transparent and robust NDFs is reiterated. The larger lesson that can be gleaned is the need for the NLP to incorporate some criteria or element about NDFs and oversight of captive breeding and artificial propagation (in addition to the Resolutions and Decisions that have tried to address the concerns with implementing these). These lessons are relevant to the next case study—South Africa—which has recently been criticized for the increasing number of captive-bred lions (William and 't Sas-Rolfes 2019), though South Africa's approach to CITES offers lessons of its own.

Case study 3—South Africa

Like Canada and Indonesia, South Africa's approach to CITES is somewhat decentralized with the provinces playing a role, particularly in enforcement. Historically, the provinces were in charge of wildlife management, but in 2004, the National Environmental Management: Biodiversity Act, Act 10 (NEMBA) came into force (DLA Piper 2015). Supporting NEMBA, which aims to conserve and manage biodiversity and protect ecosystems, including species not targeted for exploitation, are the Threatened or Protected Species Regulations 2007 (ToPS) (DLA Piper 2015) and CITES regulations (G1). I mention the exact pieces of legislation because of the somewhat complicated division of national versus provincial duties. According to one interviewee:

> The Department of Environmental Affairs (DEA) is the National CITES Management Authority who corresponds with the [CITES] Secretariat and other Parties, coordinates the implementation of CITES in South Africa and issues permits for organs of state. The nine Provincial Management Authorities issues permits for private individuals and local government bodies.
>
> (G1)

The best practice comes from the uniform guidelines for implementing ToPS and CITES between the provinces that were written by members of the national Permit and Enforcement Planning Committee (G1). In parties that have some decentralization in their implementation, uniformity and consistency are key to CITES implementation. This is to ensure that unscrupulous traders and people who commit wildlife crimes are not able to choose a province or state with the weakest legislation in which to flout or break the rules (DLA Piper 2015). DLA Piper (2015: 355), in their report on CITES legislation, judged that South Africa has "excellent laws in place". Yet, one interviewee noted that ToPS is implemented only in seven of the nine South African provinces (G1). Thus, while the legislation may be excellent, lack of implementation may impede conservation and wildlife management.

Further best practice is evident in the make-up of the authorities. "The CITES Management Authority has a member on the Scientific Authority and the CITES Enforcement Focal Point also attends the Scientific Authority meetings" (G1). Moreover, there is sharing of communications and decisions between all of the authorities and the various committees and working groups to ensure everyone is aware of ongoing activities (G1). Thus, as was evident in Canada as well, the working relationship between the authorities, in this case also three with an Enforcement Authority, seems to improve effectiveness.

I have mentioned the necessity to incorporate local and indigenous voices within wildlife management, and South Africa's approach to prepare for the CITES CoP is a good example of how this might be done.

> South Africa undertakes a public participation process in the build up to any COP. Any interested and affected parties can submit comments on any proposed uplisting or downlisting of species and then join the Management authority and the scientific authority in a meeting to discuss aspects the interested and affected parties feel the government should raise.
>
> (CS6)

As with solid legislation being hampered by the failure to fully implement it, the premise of the public participation is excellent, but the actual use of it may be limited.

> While in theory this is a step in the right direction, in practice public participation has in fact been limited. For example, in the last CoP, it was reiterated by the representative of the scientific authority that proposals brought forth by South Africa, namely on Aloe Ferrox, African Elephants and the Black Rhino hunting quota would not be discussed because South Africa had predetermined position on these proposals. This was not in the spirit of participation at all.
>
> (CS6)

Summary

It is important to ensure uniform implementation of CITES (and other regulations and legislation) when there are elements of decentralization. This addresses the possibility that criminals will cherry-pick a region or province based upon some having weaker legislation (or enforcement). Again, the value of good working relationships and clear lines of communication between authorities is evident in the structure of the CITES committees adopted in South Africa. Public participation is a critical element to CITES to ensure the public's support for decisions and thus compliance with the regulations. Such participation has to be meaningful, rather than superficial.

The case studies then provide more detailed specific recommendations for how parties might approach improvement of CITES implementation and compliance. These only form part of the picture, though, about how CITES, overall, can better protect wildlife.

Further thoughts on improving CITES

"CITES is great—best thing we've got. Implementation though at the national level is failing. Why no disciplinary action for non-compliance and not doing

NDFs [Non-detriment Findings]?" (A3). Overall, all my participants—in the Delphi iterative survey and the interviews—seemed to think CITES was helping to protect species from exploitation, but that it could do better. I would characterize my participants' attitudes towards CITES as one of resignation; they have to believe that it is doing some good, because it is the only real effort to address species loss to international demand and trade, and if CITES is not making some positive difference, then no effort is being made. The latter is too depressing to face and an admission that 45 years of hard work is not paying off.

Thus, despite numerous comments that CITES is doing a good job—and to be fair, it is likely to be helping to an important degree—the aspects for improvement are more numerous. Some of the areas for improvement, which go beyond aspects of implementation and compliance though, that are still related, are: integration of science in decision-making, holding parties to account for poor implementation and failure to comply, getting the right people in the room, and the need to modernize.

Integration of science in decision-making

As Former CITES Secretary General, Wijnstekers (2003: 80), noted in an earlier edition of *The Evolution of CITES*, "Many countries of origin lack the scientific data on the status of their animal and plant populations which makes it impossible to calculate the effects thereon of different levels of exploitation". And the most frequent suggestion for improvement was that science needed to be at the core of CITES decision-making. This came from both participants, who clearly identified themselves as being proponents of sustainable use and also from participants, who could be categorized as being protectionists. Science is possibly the most relevant when making decisions about which species need to be listed in the CITES appendices. As one interviewee stated:

> It takes too long to list and there is not enough science informing the decision. There needs to be another approach to listing such as when certain boundaries about trade numbers in combination with population size are reached, the species immediately goes into an appendix.
>
> (A3)

Other participates noted that there has been talk in the last few years of science being more integrated into decision-making processes, but this trend has gone quiet (CS5, CS8).

Several participants felt that emotions and concern for animal welfare are what block science from underpinning the decisions to list species or move them within the appendices (CS6, CS8) (which I discuss more in Chapter 7). Others felt that politics and financial interests are what keep science from being utilized (A3, CS7). Either way, in general, there is consensus that evidence of the species' population and biological characteristics, the volume of trade, and other relevant factors that impact species' survival should be driving the decisions. I suggest that this science needs to be applied to all species and not just terrestrial vertebrates.

Holding parties to account for poor implementation and failure to comply

CITES experts are well aware that the state of implementation of CITES legislation by parties could be greatly improved. While this involves actions by the parties themselves, my participants also noted that CITES could be better at holding parties to account (A3, CS7). As one interviewee said "Bans have a role to play and can be temporary" (A3). The bans they were referring to are trade sanctions on non-compliant parties. In general, interviewees and Delphi iterative survey participants expressed support for CITES adopting a stricter stance with poor implementation and non-compliance. Particularly as the approach has, historically, been more lenient, this has resulted in nearly half of the parties still having legislation that does not meet the four components of the NLP.

Making parties more accountable could also entail a return of the non-compliance report to be shared at the CoP or at a Standing Committee meeting. Internal scrutiny by fellow parties in this manner could be a useful tool to bring (some) parties into compliance as well as improve implementation. Also, making this information more visible on the CITES website, as discussed in Chapter 5, might also incentivize parties to improve compliance. The potentially increased scrutiny from members of the public, watchdog organizations, and other relevant stakeholders is another tool to bolster compliance and hold parties to account.

Getting the right people in the room

Several interviewees and Delphi iterative survey participants commented that the people with the needed expertise were not always included in decision-making processes or consultations (CS5, CS8, G5). Perhaps surprisingly, the experts who are being left out are law enforcement officials and specialist environmental or wildlife lawyers. I have seen this first-hand at both the 2018 London Conference on Illegal Wildlife Trade and the 2019 CITES CoP in Geneva, Switzerland. At the London Conference, a small minority of the keynote speakers were law enforcement officials despite the focus on crime. Furthermore, a meeting of the INTERPOL Wildlife Crime Working Group overlapped part of the event, so law enforcement officers would not have been able to or may not have chosen to attend even if their expertise had been sought. Similarly, at the CITES CoP, even though illegal wildlife trade is a core concern and, in general, management of wildlife trade is inseparable from regulation and enforcement, again law enforcement officials are in the minority in terms of participation. That is not to say that they should be dominating the discussions, but there is an argument to be made for their voices to be heard. The same could be said for specialist lawyers. With a clear lack of implementation of legislation by nearly half of the CITES parties, there is a strong case to be made for greater inclusion of legal specialists. One interviewee argued "CITES is also a multilateral environmental agreement operating in the field of international environmental law, specialist lawyers should also be more involved in this" (CS6).

Thus, while some experts are missing from the discussions, there was also indication of the inclusion of people who should not be involved in decision-making. For instance, "There is also in my view over engagement with people who have a vested financial interest in trade and limited engagement (engagement treated like a tick box exercise) for those working in the field but not trading" (CS6). In particular, the earlier examples highlight that in some instances traders and exporters are influencing the setting of quotas (A3, CS5, CS7). There is also the issue of civil society actors having too prominent a role in decision-making, which I speak about in Chapter 7.

The need to modernize

The last area for improvement is perhaps the largest and possibly the most important. CITES is 45 years old and drafted under very different political and environmental circumstances. There may well be the need for it to modernize and adopt new approaches. One reason for change is CITES' slow machinations, which means it takes years for a species to be protected.

> CITES is by its nature reactive, you have to wait for a species to be in decline before you can apply the criteria to list that species—this allows for no proactive measures to be taken to protect a species, unless you list it on App III.
>
> (CS8)

In the case of the pangolin, the eight species were already experiencing a downward spiral in population numbers, and the listing of them proposed in 2016 has done nothing to stop this. Therefore, consideration should be given to an updated approach to CITES listing that does not come too late. Perhaps something like what was mentioned earlier, that once a certain set of boundaries are reached a species is automatically added to one of the appendices, should be considered.

Part of the need to somehow automate species listings might stem from the continual growth of the international trade in wildlife and the vast number of individuals who are now traded to all parts of the world. Another feature that possibly has changed in relation to the international wildlife trade is the involvement of more powerful actors such as organized crime groups (UNODC 2020; Wyatt et al. 2020) and corporations (Wyatt et al. 2020). In regard to organized crime groups, I elaborate on how their involvement might be tackled in Chapter 7. In terms of corporations, some parties already recognize the role of big businesses in wildlife trade and trafficking and have specific parts of their legislation that addresses this. For instance, The Gambia has a specific section of its Biodiversity and Wildlife Act 2003 that implements CITES, which outlines the offences by "bodies corporate".

> Where a body corporate is guilty of an offence under this Act or any regulation made under it, every director, the secretary or any similar officer of the body corporate shall, unless he or she proves that the offence was committed

> without his or her knowledge or consent, also be liable to be prosecuted, tried, convicted and punished for that offence.
>
> (The Gambia 2003: 37)

This is a particularly interesting approach to hold individuals of companies to account for illegal behaviour. The Gambia is not alone in taking this approach; four other parties' legislations appear to have penalties that hold individuals of companies liable for the crimes of their companies. In total, it appears that 16 parties' legislation addresses corporate crime specifically. More common than the individual liability approach is to legislate for greater fines for corporations than for individual people violating CITES legislation. Addressing businesses' role in CITES offences is a key element of modernization.

Another reason to modernize CITES, as argued by one interviewee, is that the old approaches are no longer working.

> The effectiveness of trade bans is being called into question, as it may have once been effective, it no longer appears to be the case. This means that CITES may need to reinvent itself in order to be effective as an instrument for international trade in endangered species that prevents species extinction as a result of trade. That is a fundamental objective of the Convention and CITES is failing. Having a trade ban is not a measure of success.
>
> (CS8)

Yet, in the eyes of another interviewee: "CITES has an old narrative. The secretary has recently been talking up trade—CITES should not promote trade; it is a regulator not a forum to enhance trade" (G3). There may be the need to reconsider the role of the Secretariat or, as Reeve (2002) suggested, to have the Standing Committee tasked with oversight of the Secretariat.

Furthermore,

> There isn't a standardised tool for helping them [CITES] set priorities, so how do they decide which species they're going to focus on? And that's, that's huge because if you have this small staff and a huge mandate and . . . species that you could be monitoring, how do you decide which ones to invest your resources in? Do you go with the global priorities, do you go with regional priorities, do you look at foreign species, domestic species, species that have economic value?
>
> (A2)

And this perhaps gets to the crux of the problem with and for CITES. A small underfunded convention is tasked with regulating and monitoring a vast industry that has grown and is growing, since CITES was created. It may indeed be the time to modernize, reinvent, or reconsider how international wildlife trade is governed.

Summary

It is likely, according to my participants, that drafting and passing a convention like CITES would be impossible today. And without CITES, the situation of wildlife exploited by international trade may well be worse than it is. So, in spite of the flaws that are visible, there is value in the Convention and there are parties that provide ideas on how to improve existing practice. Wherever possible, there is the need for inclusivity (of stakeholders, of species), uniformity (of implementation, of legislation), consistency (of implementation, of issuing permits, of holding parties to account), and transparency (of decision-making, of holding parties to account). Clearly, this is a monumental task for such a complex global Convention. At the core of making this work is to foster good relationships wherever possible. This is relevant within authorities, between authorities, between authorities and local, regional, national, international and supranational agencies and organizations, and between the authorities and all members of the public, particularly those communities reliant on and in proximity to wildlife.

Science needs to be placed at the heart of the CITES decision-making process, and the evidence that the science produces must be widely and openly available. NDFs and all of the data that go into informing those should be on the CITES website. And the parties' engagement or not with this transparency should also be easily seen on this public forum. With limited funds to support CITES, more pressure from the parties on each other is needed to further improve CITES and truly protect the wildlife it was designed to protect.

Note

1 See Wyatt, T. 2020a. Canada and the Convention on International Trade in Endangered Species of Wild Fauna and Flora (CITES): Lessons Learned on Implementation and Compliance. *Liverpool Law Review*. DOI: 10.1007/s10991-020-09267-8 for a complete discussion of Canada's implementation of and compliance with CITES.

References

Bar-on, Y., Phillips, R. and Milo. R. 2018. The biomass distribution on Earth. *PNAS*. 115 (25): 6506–6511.

DLA Piper. 2015. *Empty Threat 2015: Does the Law Combat Illegal Wildlife Trade? A Review of Legislative and Judicial Approaches in 15 Jurisdictions*. Washington, DC: DLA Piper and WWF.

Margulies, J., Bullough, L., Hinsley, A., Ingram, D., Cowell, C., Goettsch, B., Klitgard, B., Lavorgna, A., Sinovas, P. and Phelps, J. 2019. Illegal wildlife trade and the persistence of "plant blindness". *Plants, People, Planet*. https://doi.org/10.1002/ppp3.10053.

Nijman, V. and Shepherd, C. 2015. *Adding Up the Numbers: An Investigation into Commercial Breeding of Tokay Geckos in Indonesia*. Petaling Jaya, Selangor, Malaysia: TRAFFIC.

Reeve, R. 2002. *Policing International Trade in Endangered Species: The CITES Treaty and Compliance*. London: Earthscan.

The Gambia. 2003. Biodiversity and wildlife act 2003. Available at: http://extwprlegs1.fao.org/docs/pdf/gam158129.pdf. Accessed 14 December 2020.

United Nations Food and Agricultural Organization (FAO). 2020. Establishing and linking global fish-trade information networks. Available at: www.fao.org/in-action/establishing-and-linking-global-fish-trade-information-networks/en. 16 August 2020.

United Nations Office on Drugs and Crime (UNODC). 2020. *World Wildlife Crime Report*. Vienna: UNODC.

Wijnstekers, W. 2003. *The Evolution of CITES, 7th Edition*. Geneva.

Williams, V. and 't Sas-Rolfes, M. 2019. Born captive: A survey of the lion breeding, keeping and hunting industries in South Africa. *PLoS ONE*. 14 (5): 1–31. https://doi.org/10.1371/journal. pone.0217409.

Wyatt, T. 2020a. Canada and the convention on international trade in endangered species of wild fauna and flora (CITES): Lessons learned on implementation and compliance. *Liverpool Law Review*. https://doi.org/10.1007/s10991-020-09267-8.

———. 2020b. Case study 2 — Indonesia. Available at: http://drtwyatt.weebly.com/uploads/2/4/3/1/24317071/indonesia_summary.pdf. Accessed 7 October 2020.

———. 2013. *Wildlife Trafficking: A Deconstruction of the Crime, Victims and Offenders*. Basingstoke: Palgrave Macmillan.

Wyatt, T., Friedman, K. and Hutchinson, A. Under review. Are fish wildlife? *Animal Law Review*.

Wyatt, T., van Uhm, D. and Nurse, A. 2020. Differentiating criminal networks in the illegal wildlife trade: Organised, corporate and disorganised crime. *Trends in Organized Crime*. https://doi.org/10.1007/s12117-020-09385-9.

7

THE FUTURE

The last chapter—The Future—begins with recommendations for improvement of CITES implementation and compliance. I then briefly touch on CITES' relationship with other MEAs and what lessons can be learned from these other MEAs. I, then, go on to speculate whether CITES is still fit for purpose or whether a new convention or approach needs to be taken in order to stop the decline of species. Former CITES Secretary General John Scanlon suggests that a completely new convention tackling wildlife crime in general may be needed, or the global community should consider including wildlife crime and trafficking within existing structures, such as the UN Convention on Transnational Organized Crime (UNTOC). I will explore these options through the use of primary data asking experts this very question and providing my own analysis of the advantages of these suggestions compared to the current functioning of CITES.

Then there is a section called "Unanswered Concerns". Throughout my study, a number of tensions and hurdles that impede CITES from improving as well as disrupting its functioning became highlighted. Without resolving these concerns or without the parties coming to some sort of compromise, I think the future of CITES is filled with arguments and stagnation. The concerns—corruption, politics and undue influence, North versus South, people versus conservation and the ethics of wildlife consumption—will be difficult to address, but dialogue must begin in order to better protect the wildlife CITES is meant to prevent from being exploited.

Recommendations

> The length and complexity of the Appendices makes the already difficult task of enforcement officers that much harder, and there is also very clearly insufficient implementation in some countries in relation to surveillance and the issuing of permits. The level of fines imposed on those involved in illegal

> traffic can be frustratingly inadequate to deter effectively. Moreover, securing sufficient financial resources to implement the Convention effectively is an ever-present problem. Additionally, despite efforts to improve capacity building, Management and Scientific Authorities are too often understaffed and their personnel inadequately trained and communication between Management Authorities could certainly be improved in many instances, as could the rate of submission of annual reports to the Secretariat.
>
> (Bowman et al. 2010: 533–534)

As Bowman and his colleagues note, there is a range of issues with CITES, but they can be addressed.

In Chapters 4, 5, and 6, I made specific suggestions for improvements based upon the best practice and lessons learned from my analysis of all of the parties' legislation, my Delphi iterative survey, and my 20 interviews in relation to the four components of the NLP as well as in regard to compliance. Box 7.1 highlights the recommendations for each of the NLP components that I think are most important as well as achievable.

BOX 7.1—RECOMMENDATION FOR IMPROVING THE NLP

Authorities

- Concentrated effort to get all members to have a Scientific Authority.
- Consider adding an Enforcement Authority to the Convention.

Prohibition

- Ensure all members prohibit violations of the convention in native and non-native species.

Penalties

- Consider a resolution or decision to set penalties, possibly in line with the UN guidelines that four-plus years indicates a serious crime.
- Audit that penalties are being given and to what levels and extent.

Confiscations

- Support members to widen confiscations to include assets and proceeds of crime.
- Wider dissemination of IFAW and IUCN guidelines on confiscation of live wildlife.

In addition to the targeted improvements to the four components, there are actions that, in general, could help to improve the implementation of and compliance with CITES of the parties as well as possibly better achieve the aim of the Convention. Overall, I recommend:

1 Focus on getting the science right.
2 More visibility on the CITES website in regard to non-implementation and non-compliance.
3 Disciplinary action should be considered for consistently non-compliant members.
4 Make the NLP fit for purpose and more robust (including marine and plant species).
5 Make NDFs available on the CITES website to increase transparency.
6 Encourage importers to increase scrutiny of incoming wildlife and NDFs.
7 Incorporate wildlife funds into the CITES national structures funded by the money from (increased) permit fees, captive breeding and artificial propagation licensing fees, fines, and asset seizures and forfeitures.

If parties and the CITES committees could focus on getting the science right, as I discussed in Chapter 6, this would provide the grounding for more evidence-based decisions and discussions (which I talk more about in the "Unanswered concerns" section later). Instead of a focus on training law enforcement, as tends to be the target of much of the debates regarding illegal wildlife trade, perhaps there should be a push for more biologists, more zoologists, more botanists, and so forth, who can collect and corroborate all the species' demographic data that are necessary not only for CITES, but also for the range of other conventions dealing with the biodiversity crisis. Such data underpin trade decisions and can be greatly improved as well as being more publicly available.

My second overall recommendation is to have more visibility about parties' implementation and compliance on the CITES website. At the time of writing, it is possible to uncover a list of those parties with trade sanctions and the reasons for these. However, a prominent place could prompt some of those parties to take action to address their non-compliance. I realize pointing out a nation's failings can be unpopular for political reasons. As I discuss later, CITES is often bogged down in, and too at the whim of, politics. A visible prominent list presents an opportunity to set aside the politics and highlight the facts of what information a party has not shared—the annual reports, the biennial (implementation) reports, the Reviews of Significant Trade, and the NDFs.

Even with assistance and other efforts to improve implementation and compliance, there are some parties which still are unwilling to fully engage with CITES. These parties should be sanctioned more regularly, more transparently, and equally. For those parties that are unable to fully engage, alternative solutions need to be found, such as more targeted support or direct financial assistance if possible. It would be up to the Secretariat to make this decision, but again this has to be done transparently and fairly. It is worth revisiting Reeve's (2002) suggestion that a committee for compliance be created, so that the process is more robust and transparent. The Standing Committee could then focus on other tasks, which should include oversight of the Secretariat (Reeve 2002).

At the heart of implementation of CITES is the NLP. The NLP needs to be fit for purpose and more robust. It needs to be a true reflection of the full extent of implementation—not just for terrestrial non-human animals and not just for native species. The NLP should also be expanded to consider other components that are critical to implementation, such as enforcement and corruption.

In line with other suggestions that I have made about transparency, I recommend having a database of NDFs available at the CITES website. These could be electronic forms that are filled out online and then visible to the parties and potentially the public. This way, the justification for trade and for quotas is visible, which means it can be verified and it also allows for my sixth recommendation, which is for importing parties to take a much more active role in requiring their fellow parties to implement and comply with CITES regulations. In particular, as shared with me, importing parties should not allow the import of wildlife where the exporting party has not completed a robust NDF. By having these transparently and prominently on the CITES website, this would enable the importing party to fairly quickly check the NDF and proceed or not with import.

As I discussed in Chapter 2, the CITES Secretariat and the CITES parties continually struggle with the lack of funding. Some parties try to address this by having a wildlife or conservation fund to support the costs of CITES Management. As recommended to me, international trade in wildlife is an optional activity, so if a party cannot afford to properly implement the Convention and manage the processes, they do not have to trade wildlife. If they choose to, however, why not set permit and other fees to such levels that fund the necessary management (CS7). The money from these sources could go into the wildlife fund. Other money from fines for violating CITES could also go into the fund as well as proceeds of crime from asset seizures and forfeitures, which should more regularly be used when enforcing CITES provisions. There should also be efforts made to have the remaining parties which have not ratified the Bonn Amendment for parties to contribute to the CITES budget to do so. Party contributions to the CITES budget could also be part of the more visible approach via the website on implementation and compliance.

CITES is, of course, one of several prominent MEAs. I briefly explored what further recommendations for CITES might be gleaned from the workings of other environmental conventions.

Lessons from other environmental treaties

An important aspect of CITES in recent years is, as the current Secretary General of CITES has been quoted as saying, "the importance of cooperation with other multilateral environmental agreements and intergovernmental organisations" (IISD 2019: no page). The evidence that this cooperation is ongoing and formal is, as the IISD (2019) has noted:

- CoP18 Doc.15.1, highlighting synergies with the Convention on the Conservation of Migratory Species of Wild Animals (CMS) and the UN Convention on Biological Diversity (CBD).
- Decisions 17.55 and 56. CCAMLAR (Commission for the Conservation of Antarctic Marine Living Resources).
- Resolution Conf. 12.4. (Doc.15.3) Global strategy for plant conservation.
- Doc.15.4 on cooperation with IPBES Intergovernmental Panel on Biodiversity and Ecosystem Services).
- Doc.15.6. UNESCO collaboration.

The links to these initiatives to protect ecosystems, heritage sites, and different species are critical considering the connections to sustainable trade. I also mention these efforts for collaboration because it raises the questions as to how do the other conventions approach implementation and compliance. Reeve (2002) suggested further exploration of what the other conventions were doing would be beneficial, such as Ramsar (Wetlands) and the Montreal Protocol, but I have not found substantial comparisons. Ramsar has more visible reporting, some of which is located on their website and had, at one point, around two-thirds of parties complying (Reeve 2002). The Montreal protocol's approach is to deny access to convention funds for parties that are non-compliant (Reeve 2002), which is possibly something to explore for CITES. With CITES being considered successful, it may not be surprising that there have not been robust comparisons. One interviewee suggested looking at the Nagoya Protocol (G3) (a supplement to the CBD on the Access to Genetic Resources and the Fair and Equitable Sharing of Benefits Arising from their Utilization). It is considered a failed protocol, but the approach to the theft of natural resources may offer insights (G3). Thus, other MEAs may yet provide other ways to improve CITES, but these would require extensive further research into their implementation and compliance procedures.

Furthermore, many of my participants thought CITES was doing a reasonably good job (particularly compared to other MEAs). They also commented that the alternative (nothing or a newly negotiated convention) would be much worse. Yet, the scale of improvement needed and the scope of the biodiversity crisis that is still ongoing despite CITES' 45 years of trying to regulate the trade in (some) endangered species still leads to the question whether CITES is ensuring that trade is sustainable? Is CITES doing what it is set out to do?

Is CITES fit for purpose?

CITES is a trade convention. Yet, as one interviewee noted, consumption and trade of wildlife have more far-reaching consequences than only within consumer markets and wildlife economies.

> There are impacts on ecosystems, public health, corruption—it [illegal trade] is a threat regardless of extinction, which is all CITES is worried about. Trade restrictions are a small part of the picture and getting bypassed, once

> depleted it is too late for the species. There needs to be a fundamental shift from the 1970s. The IPBES [2019] has dramatically shown that the landscape has changed for the worse. We have to prevent loss. The Convention has been good but is still going backwards.
>
> (G3)

This highlights a lack of linked-up thinking that might mean CITES needs to adapt to a global society that is perhaps more consumer-orientated and globalized than when it was drafted, as I proposed in Chapter 6 when discussing the need for CITES to modernize. Trade of wildlife, even if it is considered and properly managed as a "renewable" or "natural resource", has consequences beyond markets. As indicated in the earlier quotation, trade has impacts on ecosystems, which means on other species and their survival; there is more to wildlife's role and value than being extinct or not extinct. Furthermore, wildlife trade clearly has public health implications that we see quite starkly during the ongoing coronavirus pandemic. At the time of writing, there are over 70 million reported human cases globally and more than 1.5 million confirmed deaths (World Health Organization 2020). And the coronavirus pandemic is more far-reaching than human health. As I discussed earlier, non-human animals are susceptible to the coronavirus as well. Mink, which are farmed widely for their fur, throughout Europe in particular, are susceptible. To reiterate the scope, in Denmark, when mink tested positive for COVID-19 in November 2020 and human workers were found to have contracted COVID-19 from the mink, the Danish Government decided to cull the entire population of 17 million mink (Rasmussen 2020). The scale of loss of non-human animal life from the inextricable link between human and non-human animal health is staggering. Yet, CITES has no provisions for health, which I return to shortly.

This lack of connectedness of CITES to other social and ecological structures is evident in other ways as well. CITES is only focused on international trade, which is understandable, but is removed from the realities of both wildlife populations and consumer demand. CITES is likely flawed from the outset by seeing wildlife as the property of individual nations. Some countries are allowed to trade because "their" populations are not endangered, but this does not recognize fundamental biology or ecology that wildlife does not confine themselves to human/political borders.

> What is international? It is a useless concept for animal populations. The giraffe probably has domestic trade which is why Southern African parties claimed the listing was outside the scope of CITES. Hypocritical as CITES has/does get involved in domestic trade with the rhino and the elephant.
>
> (CS3)

The mention of the giraffe is in reference to heated debates at the CoP in August 2019 (which I attended and heard some of these debates). A proposal to list the giraffe in Appendix II was hotly contested because opposition parties claimed that there was no evidence to suggest that international trade of giraffe and/or giraffe products were threatening their population numbers. Proponents of the listing seemed to argue that while international trade may be limited, this amount could not be sustained in addition to the amount of domestic trade that was taking place. But, as I reiterated here, CITES is only concerned with international trade, so opponents of limiting the trade in giraffes and their products were saying that consideration of domestic trade was beyond the scope of CITES. I suggest this is another example of the lack of linked-up thinking or recognition of connections. Domestic trade levels surely affect species' populations, and this is certainly relevant to international trade, which cannot and does not sit alone.

In the sphere of domestic trade and demand for wildlife, there is other evidence of the lack of linked-up thinking. "Once in a country [there is] limited ability for countries to do anything about protected species" (A2). In this case, the lack of linked-up thinking originates in the domestic legislation of parties to CITES. If illegal wildlife is successfully smuggled into a party's territory, there is the possibility that nothing can be done. Legislation needs to be in place to require the correct permits as well as to prohibit the possession of wildlife without this documentation. Further disconnect at the domestic or national level stems from the failure of agencies to collaborate with one another. For instance, public health agencies (which should be aiding with disease transmission from the wildlife trade) are often not working with CITES officials (A2). In addition, in some parties, law enforcement agencies like national crime units do not have memorandums of understanding with CITES authorities to facilitate enforcement and investigations of CITES violations (G3).

"International Environmental Law is vast, so there are not many instruments like CITES that are transactional. It has a narrow brief and can be strong" (G3). It is, then, probably within the narrow brief for which it was designed that CITES is somewhat fit for purpose. That narrow brief, however, needs to be linked in and connected to the larger global mechanisms and activities that impact wildlife trade. Possibly, the main global activity which CITES is not suited to cope with is the illegal wildlife trade.

A wildlife crime convention or protocol to UNTOC?

Reeve (2002) noted nearly 20 years ago that at the centre of policing international trade in endangered species is CITES. Parties must implement and enforce the Convention in the absence of an international wildlife law enforcement agency (Reeve 2002). But, as I have documented in this book, there is no requirement for parties to enforce the Convention or to have an Enforcement Authority. Former Secretary General of CITES, John Scanlon, is vocal that a new approach to tackling illegal wildlife trade and wildlife crime is needed in order to reduce biodiversity loss from these activities. As a founding member and now chair of the Global Initiative to End Wildlife Crime (2020a, 2020b), he has been instrumental in advancing discussions about having a new convention on wildlife crime or, as the initiative has proposed, a fourth protocol to UNTOC. They have advanced this conversation so much that at the October 2020 CoP for UNTOC, the initiative held a virtual side event where they presented a draft of this fourth protocol (Global Initiative to End Wildlife Crime 2020b).

Since I am particularly interested in the illegal aspect of wildlife trade and I was aware of Scanlon's proposals, I explored in my Delphi iterative survey first round and in my interviews whether or not my participants were supportive of a new approach to tackling the illegal aspect of wildlife trade. One interviewee said to me:

> It [CITES] has exhausted its role in IWT [illegal wildlife trade], which needs to be embedded elsewhere within a part of the international criminal law system where there is infrastructure and processes to deal with transnational organized crime. Need to re-think where crime is located. It was thought CITES would never be embedded in the GEF [Global Environment Facility], but it was, so there is hope for IWT to find a place as it has more attention and its profile has shifted. There is talk about environmental crime in general being added to UNTOC.
>
> (G3)

I asked participants of Round 1 of my survey: Should there be a new international treaty or convention dealing specifically with wildlife crime and trafficking? (see Table 7.1) Most participants either disagreed (10 of 32) or neither agreed nor disagreed (11 of 32). Four strongly disagreed and only two strongly agreed that there should be a new convention. Interestingly, one participant said "Not sure if I understand the point. Isn't that the purpose of CITES"? But CITES is not designed to be a convention about policing. I also asked "Or, should wildlife crime and trafficking be added to the Protocols of the United Nations Convention on Transnational Organized Crime (UNTOC)"? (see Table 7.2) Twenty-five of 32 participants either strongly agreed or agreed. Five responded neither agree nor disagree and two participants disagreed. One respondent wrote "I have a slight preference as [for] a protocol as I believe it would be easier to achieve than a new treaty". Someone else suggested that an annex could be added to CITES.

One of the key elements of crimes that are included within the protocols of UNTOC, such as human trafficking, smuggling of migrants, and trafficking of

TABLE 7.1 Should there be a new international treaty or convention dealing specifically with wildlife crime and trafficking?

Option	*Number of Responses*
Strongly Agree	2
Agree	5
Neither	11
Disagree	10
Strongly Disagree	4

TABLE 7.2 Or, should wildlife crime and trafficking be added to the Protocols of the United Nations Convention on Transnational Organized Crime (UNTOC)?

Option	*Number of Responses*
Strongly Agree	12
Agree	13
Neither	5
Disagree	2
Strongly Disagree	0

weapons, is that the signatory parties need to recognize the offence as a "serious crime" by having a penalty of at least 4 years of imprisonment or a more serious penalty (UNODC 2000). In the new version of the implementation report, the CITES Secretariat is asking whether parties are already taking this approach. In Question 1.7.3a, parties are asked whether CITES offences can be investigated, prosecuted, and penalized criminally. Of the 34 responses, 30 answered "Yes", two "No", and two did not answer. Of those parties that have answered this question, the overwhelming majority have placed violations of CITES with the criminal justice system rather than in the administrative or civil realm, for at least some of the violations. And in Question 1.7.3b, the parties are asked "Are criminal offences such as poaching and wildlife trafficking recognized as serious crime in your country?", with serious crime referencing the UNTOC text (CITES 2018: 9). In this case, of the 34 responses, 20 answered "Yes", ten "No", and four did not provide an answer. Of the small number of respondents then, a majority have already taken the approach of UNTOC and, for at least some violations of CITES, have the ability to seek a rather strict punishment of at least four years in prison.

To me, it is clear that CITES, as it is currently structured, is not up to the task of policing illegal wildlife trade. During the aforementioned side event at the CoP for UNTOC, I spoke about the advantages for law enforcement of having a fourth protocol to UNTOC. CITES is the only international mechanism that currently attempts to address wildlife trafficking. Yet, CITES does not require that violations of the Convention be defined as crimes. So, collecting, harvesting, killing, transporting, importing, exporting, possessing, and so forth, in breach of a parties' national legislation transposing CITES is not necessarily a crime. So, in the first instance, having a fourth protocol specifically about wildlife trafficking sends a clear message about its severity. This could have deterrence effects; global recognition as one of the most serious forms of organized crime could be symbolically powerful—though this would have to be investigated.

Furthermore, as the only current mechanism for addressing wildlife trafficking, CITES is not up to the task of being a policing or a law enforcement agency for the reason I just said, but also because there are no requirements for the parties to have an Enforcement Authority, there is no obligation for the criminal justice system to be involved though obviously there is a role for regulatory agencies like border and/or customs. At present, in CITES parties without Enforcement Authorities (which is 85 parties, so nearly half of the total signatories), Ministries of Environment, Institutes of Zoology, Botanic gardens, and Fisheries Research Centres are up against criminals, who can be organized. And as I have said throughout this book, nearly half of the parties have not fully implemented the Convention as required, so the level of enforcement is nowhere near what it needs to be to begin with.

CITES is a trade treaty and not environmental or wildlife law. A fourth protocol on wildlife trafficking presents the opportunity to bring the principles of wildlife law to bear to protect species. These include, but are not limited to, prohibition of acts which harm individual animals beyond a permitted level. This is beyond the mere 38,000+ species listed in CITES, which is critical since 1 million species face extinction. Wildlife law also incorporates *res nullius* or *res communis*—implementing the public good and wildlife trust doctrines as argued by Nurse (2016).

Although the CITES Secretariat and the Standing Committee have the ability to sanction parties for non-compliance with the Convention, this does not happen all that often, even though, as I have said numerous times, nearly half of the parties have not implemented the required legislation. In addition, again as I have said before, there are numerous other issues with compliance with CITES requirements (failure to submit annual reports that include the trade data, failure to submit biennial reports that include updates on implementation (several parties have never submitted these), and failure to submit illegal wildlife trade figures (as has been required the last few years). All of these data are critical to understanding the nature and scope of wildlife trafficking and developing effective responses. It is not working to have a small staff of this Convention overseeing state obligations that are, in essence, breaching international wildlife protection standards.

And finally, in terms of defining, recognizing on an international scale wildlife trafficking as a crime, gives the global community, particularly the legal and law enforcement communities, the chance to:

- Draft an up-to-date criminal definition of wildlife trafficking that takes into account modern forms of committing crimes, like cybercrime, and including the diversity of wildlife (including plants and fish).
- Adopt measures aimed at reducing wildlife trafficking, not hindered by conflicts to uphold trade and in a proposed draft by the Global Initiative to End Wildlife Crime obligations to reduce demand, which are critical and an excellent addition.
- Promote cooperation between state parties, particularly around information sharing.
- Protect wildlife and prevent the worst forms of their exploitation.

In regard to the benefits to the criminal justice and law enforcement stakeholders of a protocol specifically, I suggest these stem from the obligations on the States. As I mentioned, wildlife trafficking under UNTOC becomes a crime in all of the jurisdictions that ratify it (which for UNTOC is more than CITES). Crimes in one nation are recognized under the proposed protocol even if they are not crimes in the nation that discovers them. Importantly, the crimes that enable wildlife trafficking also are required to be in a parties' legislation too. Money laundering then becomes a central element for parties to address as does corruption and obstruction of justice. This is likely better in terms of corruption, which is absolutely central to the facilitation of wildlife trafficking, but only featured in 2016 as a Resolution in CITES and is non-binding (which I talk about later). The required addition of these crimes extends the tools available to law enforcement officials and has the potential to operationalize them more widely.

In addition, UNTOC protocols also make more concrete the cooperation between law enforcement agencies in different countries. This stems from the requirement to have frameworks for extradition and mutual legal assistance. These more formal relationships can potentially combat corruption and possibly be more financially supported, since UNTOC has a budget that is more stable than that of CITES. There are a variety of initiatives and programmes that are supporting the training of wildlife law enforcement officials and trying to build capacity of CITES authorities, customs, forensics, and so forth. My sense from talking to those giving these trainings, and seeing the different funding streams from the US, UK, and elsewhere, is that there is a lot of replication and waste in the current system. With a singular space where wildlife trafficking is defined, there is the potential to centralize or at least use UNTOC as the means of coordinating the training and assistance that is needed to combat wildlife trafficking, likely in collaboration with INTERPOL's Wildlife Crime Working Group. It will be interesting to watch, though, how the Global Initiative to End Wildlife Crime's efforts for a fourth protocol evolve.

Unanswered concerns

For the last section of this book, I am going to stray beyond the core themes of implementation of and compliance with CITES. Over the last several years, as I have immersed myself in the workings of the Convention, it has reiterated to me that implementation and compliance are much more complex than simply having the correct legislation and completing the correct reports on time. There are numerous social and political elements to the global wildlife economy that affect whether or not parties are able or willing to fully engage with CITES. With that in mind, I talk about my unanswered concerns—those social and political elements which I do not see immediate resolutions to, but which need to be addressed to improve CITES and to stem the biodiversity crisis linked to the trade and trafficking of wildlife.

Corruption

One of the biggest challenges for CITES is corruption. This can impact on the implementation of CITES if the political, legal, and criminal justice systems in a party are compromised (Wyatt et al. 2017). Corruption can also greatly hinder the permit system as well as enforcement of violations and the resulting sanctions. CITES is well aware of the significant role corruption plays in the illegal wildlife trade, and in 2016 there was a Resolution addressing it. Resolution Conf. 17.6—Prohibiting, preventing, detecting and countering corruption, which facilitates activities conducted in violation of the Convention—starts by saying "AWARE that corruption can play a significant role in facilitating activities conducted in violation of the Convention at all points of the trade chain, in source, transit and market countries" (CITES 2016: 1). Since this is a resolution, it is only guidance. As one interviewee stated:

> [The] Resolution on corruption is great in principle, but as a resolution is toothless. It is a desperate issue, but unless a decision is taken, which can't happen until 2022 CoP, really no changes improvements will take place. Governments need to put forward a decision, or the Secretariat needs to follow up somehow with the resolution. Donors need to build in corruption accountability into their funding.
>
> (CS7)

A concrete effort to map the legal provisions related to corruption for each of the parties is warranted (and I did hear speculation that this was taking place though I have not been able to verify this). And since corruption is such an obstacle for CITES, it should be considered for corruption to be an annex to the Convention and integrated into the components of the NLP.

Politics and undue influence

In my recommendations for overall improvement, I proposed "focusing on the science". This was repeated to me over and over again in the interviews. It is in contrast to the perceptions of many of my participants of the current functioning—that CITES is about the politics of wildlife trade. It was explained to me that the decision-making process is not about the conservation and survival status of wildlife, but rather about the economic interests of some parties or, alternatively, about the prioritizing of animal welfare and animal rights (which I return to in the people versus conservation and ethics sections in a moment). Listings are often made for political, economic, philosophical, and even emotional reasons, as well as scientific reasons (Vincent et al. 2013; Webb 2013).

The lengths that some parties or stakeholders go to secure votes in favour of their position are eye-opening. "There is far too much politics. Sometimes blatant bribing with passing of envelopes before votes can be seen at CITES CoP plenaries" (CS2). Another interviewee also mentioned the bribery for votes (A1). In addition to rooting out corruption that impacts CITES implementation, as I outlined in the earlier section, there should be initiatives to squash corruption in the activities of CITES actors as well. Several of my participants pointed to civil society or non-governmental organizations (NGOs) in this context. NGOs have a huge presence at the CITES CoP and are responsible for so much stuff going on behind the scenes (A1).

Challender and MacMillan (2019: 91) have noted

> (i) by design, non-state actors have played a role in CITES proceedings since the Convention's inception in 1975; (ii) although CITES is concerned primarily with sustainability in international wildlife trade, the issues it addresses are inherently political, economic, and cultural, as well as emotive to some actors; and (iii) the conduct of non-state actors has recently been called into question by some CITES parties concerned that nongovernmental organisations (NGOs) are "overstepping the mark", resulting in calls for a code of responsibility.

Several interviewees referred to the lobbying power of NGOs.

> Under the influence of wealthy NGO lobbyists, however, some CITES parties do not always use evidence to make decisions and appear to be swayed by emotion and money, and CITES has, therefore, lost some credibility with sustainable use countries. CITES needs to better integrate evidence into their decision-making process and reduce the influence of lobbyists at their CoPs. (CS6)

Other interviewees disagreed that NGOs lobbying was an issue. "NGOs lobbying power isn't a vote. If you want to sway voters, lobby too" (G3). Or, as another participant said, "The beauty of CITES is that everyone gets one vote in order to govern the global commons. NGOs may testify and provide evidence, etc., but they aren't also voting members—one member one vote" (CS7). Whereas it is true that CITES operates with one vote per party, as Sollund and Runhovde (2020) point out, this does not make for an equal field. The more well-resourced parties, which are mostly in the global North, send large delegations to the CoP. These delegate members can sway the decisions made through lobbying efforts partly because they have the capacity to be more informed about the Resolutions, Decisions, and listings being proposed (Sollund and Runhovde 2020). The same can be said for the NGOs. One interviewee told me that the World Wildlife Fund sent several dozen staff to the CoP in Geneva in August 2019. This number of people is much more than many official parties can send. It also costs for most stakeholders to attend the CoP (USD 600 for the first delegate, and USD 300 for each additional person (CITES 2019)), so only well-resourced parties or NGOs can afford to send such large delegations. This disparity in ability to fully participate brings me to my next concern regarding the tension between the global North and global South.

North versus South

The divide between the North and South at CITES appears to stem from a range of sources. First, as I mentioned in the previous section, there is the disparity in resources (both to implement and comply with CITES, but also to engage with CITES through committees and the CoP). In general, global North stakeholders are the more well-resourced actors compared to their counterparts in the global South. This difference leads to feelings and perceptions of the more well-resourced parties controlling the direction of the Convention and the decision-making. Second, I was told:

> The Southern African faction feels sidelined as they can be out voted with the EU countries voting as a block. The Global South can be silenced with this approach to voting. They feel they don't have the same representation and the Global North are making the decisions.
>
> (A1)

Since the European Union has adopted a system of voting altogether, this holds significant sway not only in terms of influence, but also in terms of sheer number of parties voting in a particular manner. The counter argument could be that other regions of the world could make similar, though less formal, arrangements to counterbalance the EU.

Lastly, there is tension about the flow of the wildlife. "[Range countries] carry a very heavy burden as custodians of irreplaceable magnificent species, while the market countries (illicit or otherwise) take no responsibility and are not held to account for the damage that they cause elsewhere" (CS8). While not all trade will involve removal of wildlife from the South and consumption in the North, this often is the direction of flow for many wildlife. The EU and the US are two of the largest wildlife consumers, and, as this participant suggests, they do very little to take responsibility for that consumption or the damage that it causes. Instead, their efforts are to somehow improve management or prevent illegality in the source countries as well as focusing on demand reduction in other areas, like China and Southeast Asia, when they need to focus on their own borders. The next two sections—People versus conservation and the Ethics of wildlife consumption—intertwine with the tensions between the global North and global South, but require more discussion, so are covered in more detail in the next sections.

People versus conservation

One of the core unanswered concerns for me relates to what I mentioned in the Introduction about utilization versus protection or sustainable use versus preservation or some combination of these terms. Often this tension manifests as the proponents of utilization are seen to be for "people". The proponents of protection are seen to be for conservation, or they are seen to care more about wildlife than they care about people. The reality is much messier, and there are variations and gradations of these, at times, nearly polar opposite approaches. The proponents of a more people- or community-focused approach to CITES argue that the needs and voices of people living with wildlife have largely been ignored when designing conservation strategies (see Cooney and Abensperg-Traun 2013 among others). Leaving local and indigenous people out of the decision-making hurts people's livelihoods as well as having a negative impact on the conservation of the species the strategies are trying to protect (Cooney and Abensperg-Traun 2013). As a trade treaty and not a conservation treaty, it may seem that having people at the heart of the decisions would make sense. However, one interviewee said to me:

> Efforts to have rural, local or indigenous communities consulted or have quotas for them to be delegate members is an underhanded attempt to force socio-economic considerations into the listing process. If a species is threatened by trade, it is listed as Appendix I or II. This needs to be based on biological/scientific evidence which links to the socio-economic since scarcity would mean reliant communities are negatively affected.
>
> (CS1)

It is interesting that both sides accuse each other of not making scientific decisions. But despite this particular participant's opposition to such consultations or quotas, initiatives are progressing to have local and indigenous voices more prominent at CITES.

> Local communities in South Africa are playing a leading role in establishing a CITES committee for voices of the people who live with wildlife. On par with the Animal, Plants and Standing Committee, which are remnants of fortress conservation that conservation is apart from humans. Communities must be integrated into conservation efforts. Communities are getting more attention at CITES, which was noticeable in 2016 Johannesburg.
>
> (A1)

The fortress conservation that this participant is referring to is sometimes an element of the preservation or protection approach. In these strategies, land is cordoned off from having people living on it or accessing the resources. National parks and other such preserves can sometimes be part of this approach. Formation of some of these areas has included forced removal of local and indigenous people from the land suddenly designated as a protected area (see Brockington 2002 among others).

I am certainly not alone in my concern about this tension.

> Protection versus utilisation remains a core tension of CITES. The giraffe at CoP18 is the perfect example. Conservationists split over the science/evidence of international trade in giraffe parts. Listing in Appendix II ensures

> trade should not contribute further to decline, but Southern African parties opposed.
>
> (CS4)

I referred to the giraffe listing earlier in the section asking if CITES was fit for purpose. Proponents of listing seem to regard that the possibility of domestic trade was relevant to the listing of the giraffe internationally because the species cannot survive the strain of both demands. Some range states, though, felt this would negatively impact the livelihoods of people trading in giraffe parts.

As I said earlier, the reality, though, is messy and there are no such clear divisions of approach, even in the African context, which can sometimes be the portrayal in the media.

> Protection over utilization in the African context is not a Western imperialist dynamic as some commentators and the media may have people believing as there are huge disagreements within Africa, some of who support bans on trade of these species. The media paints bans and trade restrictions as a threat to the livelihoods of local people, but fail to cover the fact that local people are those hurt most by poaching. Southern African members are not budging on this issue unlike Japan, who, after years of trying to get the minke whale downlisted, gave up as the rest of the world wasn't moving in that direction.
>
> (CS7)

Others, though, point out that the resistance by some range states is linked to Western colonization and imperialism.

> The dissatisfaction or vitriol of Southern African countries is not a new thing. The sustainable development paradigm is in parts of Southern Africa. There is a sense by range states that they are rendered voiceless and have no say about their own animals, why are their animals' global commons and not property, what about sovereignty? There is a lot of neo-colonial posturing. Environmental affairs and ministries have limited funds. There are limited support and funds for reserves and national parks, so sustainable use is the only viable option. It is difficult for people living with the wildlife to understand why a few can't be hunted to pay for conservation and to have industries in these communities supporting people.
>
> (A1)

The tension between people versus conservation links to my discussion of whether or not CITES is fit for purpose, when it is an international treaty or approach trying to manage local people's engagement with wildlife in their countries. As one interviewee noted: "[There are] Opposing opinions and depends on local wildlife populations. Global populations declining [of wildlife] are hard to justify/explain in light of local experiences" (A1). And for those advocating for utilization: "The global community often feels like a lonely place for a country like South Africa in the CITES and sustainable use context" (CS8).

Some opposition to sustainable use is not based upon an animal welfare or animal rights ideology (that I talk about in the next section). Instead, one participant told me:

> Small-scale wildlife trade was thought to be the saviour of wildlife, but it isn't competitive with land use as the income is small, so species numbers are decreasing. A decent level of trade that is ecologically and financially sustainable is not easy. In Indonesia, people are surviving off of catching wild pythons, but there are only that many pythons because they live in the waterways of the palm oil plantations—how is it that conservation?
>
> (G3)

Furthermore, "There are conflicts with communities though [regarding sustainable use] and the big push for captive breeding has not benefited them only rich individuals and corporations. Most exports of wildlife are captive-bred, but this doesn't benefit locals" (G3). Thus, some detractors from sustainable use argue that it does not provide the livelihoods and benefits that proponents claim that it does. In addition, in some places, the conditions to support sustainable use are detrimental to the environment, which adds further negative environmental and social consequences.

It seems though that many proponents of sustainable use believe that opposition to their stance is based in animal welfare and animal rights.

> We believe that CITES should encourage sustainable use of CITES-listed species. Only species that fulfil the listing criteria of CITES should be listed on the Appendices. It seems that CITES decisions are no longer based on scientific evidence but rather on emotions and animal rights.
>
> (G1)

Two points about this interviewee's comment. First, there seems to be a fundamental misunderstanding about the role of CITES. CITES is there to guarantee the survival of the species who are being traded. They are not an advocacy group or industry representative that should be promoting trade of the species listed. Second, the remark about decisions being made based upon emotions (and animal rights) fails to recognize that their stance is clearly grounded in strong emotions as well. The "people" side of the people versus conservation debate is equally emotive.

This tension is not one that is going away:

> Every CoP some issue of utilization versus protection comes up and southern African countries are nearly always out of step with other members and

> particularly other members in Africa. More countries each time are on the side of protection when it comes to giraffes, elephants, rhinos.
>
> (CS7)

A final point regarding people versus conservation is that this is not confined to Africa, nor to just the physical global South, as highlighted by this excerpt.

> TERRY AUDLA, a Canadian Inuit leader, grew up in the remote Arctic town of Resolute Bay, where children used to huddle near the school door during breaks for fear of prowling polar bears. The Inuit, he says, see the animal as a fierce predator, a cultural symbol and a valuable source of food, warmth and money in a part of the world where all three are in short supply.
>
> Yet to animal-welfare and green groups in warmer places the polar bears are both an icon in the fight against climate change and an animal under threat of extinction. The melting of the Arctic's ice cap, which the bears use as a hunting platform, means the estimated population of between 20,000 and 25,000 will decline sharply, they say. They see hunting the bears as an anachronism and want international trade in bear pelts and parts, already severely restricted, completely banned.
>
> (The Economist 2013)

The tension between utilization and protection—people versus conservation—is everywhere, and I suggest that at the heart of this tension there is an ethical debate about humans' relationship with nature and with other animals in particular.

The ethics of the wildlife trade

For me, there are two separate ethical dilemmas evident in the ongoing debates and actions of the parties to CITES. First is that CITES does not afford the same consideration to all species who are endangered and who are not being traded at sustainable levels. Second is a much more fundamental and moral question. With all that we now know about the importance, sentience, and lives of non-human animals in particular, but wildlife in general, is it ethical to continue consuming them in the way that humans do or at all?

In regard to unequal consideration between species, there is the issue of listings and then the issue of the legislation itself. In regard to listings, studies have shown this speciesist bias in the listings of CITES. Both Fukushima et al. (2020) and Scheffers et al. (2019) document that there are numerous species threatened, at least in part, by trade and are not listed in the CITES appendices. Invertebrates such as corals, sea urchins, and sea cucumber, which are particularly under pressure for commercial exploitation, are likely to be underrepresented in CITES (Fukushima et al. 2020). Hutchinson et al. (Under review) document the same speciesism and show that what seems to be driving this bias is not just human preference for the charismatic megafauna like elephants and rhinos, but also prioritizing the continued commercial exploitation of the most valuable species regardless of the threat to their survival. This seems most evident in the failure to protect commercial fish species on the CITES appendices (Wyatt et al. Under review). Further evidence comes from an analysis of those species for whom parties have placed reservations. Reservations for a CITES-listed species means that the party placing the reservation carries on trading the wildlife as if the species were not listed or as if the party was not a member of CITES. The species under reservations currently are predominantly furbearing mammals (foxes and weasels), which are valued in the fur industry (Hutchinson et al. Under review).

Some of my Delphi iterative survey participants and interviewees also suggested CITES has speciesist tendencies. One survey participant wrote: "Not protecting all species—need to include fisheries agencies in the authorities". Another survey respondent noted: "Current feedback results in heavily weighted anecdotes and focus on charismatic species from developing countries". As I mentioned when detailing the NLP, one interviewee proposed that the NLP categories, as they stand, are only part of the implementation picture, because the assessment does not include implementation of the four components for marine species (CS7). Thus, not only do fewer species that need protection get listed on the CITES appendices, if they are not charismatic and they have too much commercial value, but even once listed these uncharismatic species are overlooked. As I discussed in Chapter 6 under the section "Legal definitions of wildlife", the core legislation where what wildlife is defined as might be a source of speciesism as well. Fish are sometimes not included in definitions of wildlife and thus, are potentially unprotected despite being listed by CITES. Much more research and improvements need to be made in this area.

This is not just the case for non-human animals. As was evident in Chapter 2 with the breakdown of species listed in CITES, in terms of plants the vast majority of those offered protection are orchids. And plants, like fish, are not always included in legal definitions of wildlife. Yet, the equally endangered, but more commercially exploited and valuable trees are less likely to be listed. This has changed in the last few years with the listing of rosewood, for instance. And it seems CITES maybe changing somewhat in its speciesism, at least in regard to giving consideration to who is already listed.

> Plants are getting more focus, such as rare succulents. Communities propagating them which helps conservation, but obviously returns to the problems with commodification of wildlife. [And the] Using of indigenous knowledge to propagate.
>
> (A1)

This comment raises my second ethical dilemma about the commodification of wildlife. The comment also brings to light a very important point around whether initiatives around sustainable use are, possibly inadvertently, further exploiting indigenous peoples by appropriating their knowledge of propagation, another element of the tensions between the global North and South.

Since CITES came into force in 1975, there has been, I do not think it would be an exaggeration to say, a huge shift for many groups of people in their regard for non-human animals and the environment. In part, this has been driven by further scientific evidence of the sentient nature of non-human animals, their rich emotional lives, and the social and familial bonds that they form (see Nurse and Wyatt 2020 among others). There are now glimmers of similar, albeit small, changes to people's regard for plants as more evidence comes to light of the vast communication networks of plants, trees, and fungi (Simard 2012; Wohlleben 2016). The continued exploitation of wildlife then raises not only the issue of non-human animal (and plant) welfare, but also the entire premise of commodifying sentient beings.

> Animal rights and welfare are central narratives now that countries opposed to need to realize and adapt to. Some NGOs are influencers, but often as a result of their constituencies, particularly for animal welfare and animal rights. Live elephants, trophy hunting—animal rights is always now in the background. There are provisions for animal welfare during trade and countries dismiss this as irrelevant, but it is coming into the central narrative. CITES is limited in what it can do for animal welfare, but it has a role to play considering the absence of other places. For sentient species, CITES is the logical place to address welfare, but it wasn't set up for this and would have to be amended.
>
> (G3)

This interviewee mentioned live elephant trade. As I mentioned, the call to ban the trade of live baby elephants was one of the possibly most hotly contested proposed amendments at the 18th CoP in August of 2019. Another interviewee agreed that this proposal, which was successful, was more about welfare than anything else (CS7). From what is known about the emotional lives of elephants and the centrality of their family and social groups to their lives, it should not be a surprise that some CITES parties are advocating for the practice of taking baby elephants from their mothers and herds to stop.

On one side, there are parties which continue to believe wildlife is a commodity; that wildlife is a natural resource to be used as needed to support human livelihoods and to meet humans' needs and desires for food, medicines, companion animals, clothes, luxury products, and so forth. Often, the commercialization and commodification of wildlife lives are prioritized over their survival and even more often over their welfare and freedom from suffering. On the other side, there are parties which, at the very least, may support continued trade, but want tougher measures to ensure the welfare of the wildlife. This is clear in the support work being done by IFAW and IUCN (2019) in producing the "Guidelines for the confiscation of live wildlife". And there are those on this side who are questioning the very activity and supposed necessity of wildlife exploitation. These are not only the more radical NGOs, like World Animal Protection (2020), but also parties which have legislation that takes a much more compassionate approach to wildlife. For instance, The Gambia's (2003: 48) Biodiversity and Wildlife Act, which states:

> The overall goal of the biodiversity and wildlife policy is to create a society that sees itself as an integral part of nature, recognizes different life forms, sustainably uses natural resources and maintains for posterity a nurturing and dynamic world rich in biodiversity.

Costa Rica (2013: 4), also has legislation with a less instrumentalized, commodified view of wildlife as well as legislation which recognizes the necessity for equality amongst people:

1 Respect for all forms of life. All the living things have the right to live, independently of actual or potential economic value.
2 The components of biodiversity are valuable. They have decisive and strategic importance for the development of the country and are indispensable for the domestic, economic, social, cultural and aesthetic use of its inhabitants.
3 Respect for cultural diversity. The diversity of cultural practices and associated knowledge of the components of biodiversity should be respected and promoted, in conformity with national and international legal standards, particularly in the case of the peasant communities, the indigenous people and other cultural groups.
4 Intra and inter-generational equity. The State and private individuals will watch over the sustainable utilization of the components of biodiversity to ensure that the possibilities, opportunities and benefits of their use will be guaranteed in an equitable manner for all sectors of society and will satisfy the needs of future generations.

The respect for all life and the call for intra- and inter-generational equity make this legislation a perhaps useful example of trying to bridge the tension between people versus conservation as well as the ethics of wildlife consumption.

Clearly, there is no easy or quick resolution to this ethical divide. And, unfortunately, I do not have an answer to this concern, or to my other unanswered concerns about the tensions between the North and South, between people and conservation, or what to do about corruption and the political nature of CITES. I can only suggest that it is passed time to have open, difficult conversations about all of these issues. At the next CoP in 2022 let us hope that each of these topics are side events at the least or timetabled topics of discussion. Without facing up to these tensions, CITES will continue to be a bitter battleground where the wildlife loses out on being protected.

Final thoughts

I find myself in agreement with many of my participants that the alternative of having no wildlife trade regulations/treaty or trying to start negotiating a new convention is worse than the current conditions. Yet, I am disheartened that there is so much to fix in the current system and possibly very little will or resources to improve it. After 45 years, there are still nearly half of CITES parties that have not fully implemented the Convention. A vast majority of them have the required Management and Scientific Authorities, but the other three components of the NLP to monitor implementation have poor uptake. Many parties have not prohibited violations of the Convention, or if they have, this is only of their native species and not all of the CITES-listed species. Some parties have no apparent or very minimal penalties. Confiscation of wildlife and wildlife products is a patchwork across the parties, leaving places that are ideal for criminals to take advantage of. Annual and biennial (implementation) reports are late or never completed. The Reviews of Significant Trade and NDFs that underpin trade decisions and species listings are inadequate. Those parties not compliant are rarely held to account. Many of these issues could be improved, and I suggest that this book provides ample possible concrete steps for such improvement, if there were the political will and resources. Yet unfortunately, the political will and resources have likely decreased even further as I write this book because of the coronavirus pandemic, which continues to ravage many parts of the world.

But, maybe, hopefully, the fact that the coronavirus has come from wildlife trade will instead spark government interest in updating CITES. The other push from the Global Initiative to End Wildlife Crime (2020b) is the proposal that CITES should amend its functioning to incorporate consideration of public health stemming from zoonotic disease. The suggestion is that CITES is the logical place to incorporate checks on non-human animal health in trade. This has promise but CITES is limited. First, it is limited in overall scope in terms of the number of total species traded—again, only 38,000+, when tens of thousands of more, at least, are being traded. Second, CITES does not address domestic demand and trade. Thus, oversight for zoonotic disease in the CITES context would touch only a portion of species that may pose a zoonotic disease threat. The efforts and conversations are ongoing, and presumably will continue, as to how best to prevent the next pandemic.

Of course, there are suggestions to drastically curtail or even end wildlife trade altogether. With the alarming state of species survival and declining biodiversity, in addition to the coronavirus pandemic, these suggestions should not be ignored just because they may appear too radical. There is a desperate, urgent need to reconsider and reconfigure many of the world's people's relationship with nature and wildlife. And this returns the debate to my unanswered concerns—in particular people versus conservation and the ethics of wildlife consumption. All the people involved in CITES, and the wider wildlife economy, need to start the difficult conversations about non-human animal welfare and the commodification of wildlife.

Wildlife trade should not be preserved at the cost of species suffering or survival. Protecting human livelihoods in the short term may not only lead to eventual species extinction, but also to lost livelihoods for future generations of people. Finding the balance between people and conservation truly means a balance, where one is not prioritized over the other. We do not, in the contemporary context with complex globalized flows of wildlife, know what that balance looks like in the hugely diverse geographies and markets of wildlife. We do, however, have a framework—the Convention on International Trade in Endangered Species of Wild Fauna and Flora—that, if improved with better implementation of existing provisions and with better compliance to what has been agreed to by a vast majority of the world's governments, could get society closer to finding that balance. With these improvements and further changes that I have set out here, CITES can still contribute to stemming biodiversity loss; it is time for concrete changes and improvements to be made in order to do so.

References

Bowman, M., Davies, P. and Redgwell, C. 2010. *Lyster's International Wildlife Law*. Cambridge: Cambridge University Press.

Brockington, D. 2002. *Fortress Conservation: The Preservation of the Mkomazi Game Reserve, Tanzania*. Melton: James Currey Publishers.

Challender, D. and MacMillan, D. 2019. Investigating the influence of non-state actors on amendments to the CITES appendices. *Journal of International Wildlife Law & Policy*. 22 (2): 90–114. https://doi.org/10.1080/13880292.2019.1638549.

Convention on the International Trade in Endangered Species of Wild Fauna and Flora (CITES). 2019. Financing and the costed programme of work for the Secretariat for the triennium 2020–2022. Available at: https://cites.org/sites/default/files/document/E-Res-18-01.pdf. Accessed 10 December 2020.

———. 2018. Implementation report 2015–2017 Austria. Available at: https://cites.org/sites/default/files/reports/15-17Austria.pdf. Accessed 15 December 2020.

———. 2016. Resolution Conf. 17.6— Prohibiting, preventing, detecting and countering corruption, which facilitates activities conducted in violation of the convention. Available at: www.cites.org/sites/default/files/document/E-Res-17-06_0.pdf. Accessed 17 November 2020.

Cooney, R. and Abensperg-Traun, M. 2013. Raising local community voices: CITES, livelihoods and sustainable use. *Review of European, Comparative and International Environmental Law*. Special Issue—Focus on: CITES+40. 22 (3): 301–310.

Costa Rica. 2013. Biodiversity law: The legislative assembly of the republic of Costa Rica. Available at: www.faolex.fao.org/document/texts/biodiversitylaw. Accessed 15 December 2020.

Fukushima, C., Mammola, S. and Cardoso, P. 2020. Global wildlife trade permeates the tree of life. *Biological Conservation*. 247: 180503. https://doi.org/10.1016/j.biocon.2020.108503.

Global Initiative to End Wildlife Crime. 2020a. Form and content of a possible Protocol on the illicit trafficking of wildlife. Available at: https://endwildlifecrime.org/untoc-wildlife-protocol/. Accessed 17 November 2020.

———. 2020b. International coalition urges countries to #EndWildlifeCrime UN world wildlife day 2020. Available at: https://endwildlifecrime.org/wp-content/uploads/2020/07/PR-Wildlife-Crime_030320.pdf. Accessed 17 November 2020.

Hutchinson, A., Stephens Griffin, N. and Wyatt, T. Under review. Speciesism and the wildlife trade: Who gets listed, downlisted and uplisted in CITES? *International Criminology*.

Intergovernmental Science-Policy Platform on Biodiversity and Ecosystem Services (IPBES). 2019. IPBES global assessment summary for policymakers. Available at: www.ipbes.net/sites/default/files/downloads/summary_for_policymakers_ipbes_global_assessment.pdf. Accessed 7 May 2019.

International Fund for Animal Welfare (IFAW) and International Union for the Conservation of Nature (IUCN). 2019. *Guidelines for the Confiscation of Live Wildlife*. Gland, Switzerland: IUCN.

International Institute for Sustainable Development (IISD). 2019. Earth negotiations bulletin. August. Available at: http://enb.iisd.org/vol21/enb21101e.html?utm_medium=email&utm_campaign=ENB%20Update%20%2030%20August%202019%20SW&utm_content=ENB%20Update%20%2030%20August%202019%20SW+CID_a7fade43500fa9f1ca5d10a7a4d2b3fa&utm_source=cm&utm_term=Read. Accessed 21 October 2019.

Nurse, A. 2016. Beyond the property debate: Animal welfare as a public good. *Contemporary Justice Review*. 19 Issue, Animals, Justice and the Law Part 2: 174–187.

Nurse, A. and Wyatt, T. 2020. *Wildlife Criminology*. Bristol: Policy Press.

Rasmussen, S. 2020. Denmark to cull 17 million mink after coronavirus mutation found. *The Wall Street Journal*. Available at: www.wsj.com/articles/denmark-to-cull-17-million-mink-after-coronavirus-mutation-found-11604592156. Accessed 17 November 2020.

Reeve, R. 2002. *Policing International Trade in Endangered Species: The CITES Treaty and Compliance*. London: Earthscan.

Scheffers, B., Oliveira, B., Lamb, I. and Edwards, D. 2019. Global wildlife trade across the tree of life. *Science*. 366 (6461): 71–76. https://doi.org/10.1126/science.aav5327.

Simard, S. 2012. Mycorrhizal networks: Mechanisms, ecology and modeling. *Fungal Biology Reviews*. 26: 39–60.

Sollund, R. and Runhovde, S. 2020. Responses to wildlife crime in post-colonial times. Who fares best? *The British Journal of Criminology*. 60 (4): 1014–1033. https://doi.org/10.1093/bjc/azaa005.

The Economist. 2013. Polar bear politics: The Inuit take on world's greens and animal rightists Available at: www.economist.com/the-americas/2013/02/23/polar-bear-politics. Accessed 19 February 2020.

The Gambia. 2003. Biodiversity and wildlife act 2003. Available at: http://extwprlegs1.fao.org/docs/pdf/gam158129.pdf. Accessed 15 December 2020.

United Nations Office on Drugs and Crime (UNODC). 2000. United nations convention against transnational organized crime definition. Available at: www.unodc.org/unodc/en/organized-crime/intro/UNTOC.html. Accessed 2 November 2020.

Vincent, A. C. J., Sadovy de Mitcheson, Y., Fowler, S. L. and Lieberman, S. 2013. The role of CITES in the conservation of marine fishes subject to international trade. *Fish*. 14: 1–28.

Webb, G. J. W. 2013. *Wildlife Conservation: In the Belly of the Beast*. Darwin, NT, Australia: Charles Darwin University Press.

Wohlleben, P. 2016. *The Hidden Life of Trees: What They Feel, How They Communicate-Discoveries from a Secret World*. Vancouver: Greystone Books.

World Animal Protection. 2020. End the global wildlife trade. *Forever*. Available at: www.worldanimalprotection.org/take-action/end-global-wildlife-trade-forever. Accessed 10 December 2020.

World Health Organization. 2020. Coronavirus pandemic 2019. Available at: www.who.int/emergencies/diseases/novel-coronavirus-2019. Accessed 17 November 2020.

Wyatt, T. Friedman, K. and Hutchinson, A. Under review. Are fish wild? *Animal Law Review*.

Wyatt, T., Johnson, K., Hunter, L., George, R. and Gunter, R. 2017. Corruption and wildlife trafficking: Three case studies involving Asia. *Asian Journal of Criminology*. 13: 35–55. https://doi.org/10.1007/s11417-017-9255-8.

REFERENCES

Aguilar, S. 2013. Regulatory tools for the management of fish and timber species through CITES. *Review of European Community & International Environmental Law*. 22 (3): 281–290.

Arroyo-Quiroz, I. and Wyatt, T. 2019. Wildlife trafficking between the European Union and Mexico. *International Journal for Crime, Justice and Social Democracy*. 8 (3): 23–37. https://doi.org/10.5204/ijcjsd.v8i3.1243.

Australian Government. 2019. Environment protection and biodiversity act 1999. Available at: www.legislation.gov.au/Details/C2019C00275. Accessed 10 August 2020.

Bar-on, Y., Phillips, R. and Milo. R. 2018. The biomass distribution on Earth. *PNAS*. 115 (25): 6506–6511.

Baynes, C. 2020. Humans told to socially distance around wild animals to avoid 'considerable risk' of passing on coronavirus. *The Independent*. Available at: www.independent.co.uk/news/science/coronavirus-wild-animal-transmission-covid-19-social-distancing-b864357.html. Accessed 10 December 2020.

BBC. 2020. Endangered tigers making a 'remarkable' comeback. Available at: www.bbc.co.uk/news/newsbeat-53581028. Accessed 10 December 2020.

Bowman, M. 2013. A tale of two CITES: Divergent perspectives upon the effectiveness of the wildlife trade convention. *Review of European Community & International Environmental Law*. 22 (3): 228–238.

Bowman, M., Davies, P. and Redgwell, C. 2010. *Lyster's International Wildlife Law*. Cambridge: Cambridge University Press.

Brockington, D. 2002. *Fortress Conservation: The Preservation of the Mkomazi Game Reserve, Tanzania*. Melton: James Currey Publishers.

Carson, R. 1962. *Silent Spring*. New York: Penguin Books.

Centers for Disease Control and Prevention. 2020. COVID-19 and animals. Available at: www.cdc.gov/coronavirus/2019-ncov/daily-life-coping/animals.html. Accessed 10 December 2020.

Challender, D. and MacMillan, D. 2019. Investigating the influence of non-state actors on amendments to the CITES appendices. *Journal of International Wildlife Law & Policy*. 22 (2): 90–114. https://doi.org/10.1080/13880292.2019.1638549.

Convention on International Trade in Endangered Species of Wild Fauna and Flora (CITES). No date a. 14.81 Great whales. Available at: www.cites.org/eng/dec/valid17/220.7 July 2020.

———. No date b. Decisions of the conference of the parties to CITES in effect after the 18th meeting. Available at: www.cites.org/eng/dec/index.php. 7 July 2020.

———. No date c. CITES compliance procedures. Available at: www.cites.org/eng/prog/compliance. Accessed 3 September 2020.

———. No date d. The CITES species. Available at: www.cites.org/eng/disc/species.php. Accessed 3 July 2020.

———. 2020a. *Notification to the Parties No. 2020/006*. Geneva: CITES.

———. 2020b. Resolutions of the conference of the parties in effect after the 18th meeting. Available at: www.cites.org/eng/res/index.php. Accessed 7 July 2020.

———. 2019a. 18.125–18.128 Electronic systems and information technologies. Available at: www.cites.org/eng/dec/valid17/82172. Accessed 11 August 2020.

———. 2019b. Conf. 14.3* (Rev. CoP 18). CITES compliance procedures. Available at: www.cites.org/sites/default/files/document/E-Res-14-03-R18.pdf. Accessed 17 July 2020.

———. 2019c. Countries currently subject to a recommendation to suspend trade. Available at: https://cites.org/eng/resources/ref/suspend.php. Accessed 26 August 2020.

———. 2019d. Financing and the costed programme of work for the secretariat for the triennium 2020–2022. Available at: https://cites.org/sites/default/files/document/E-Res-18-01.pdf. Accessed 10 December 2020.

———. 2018a. Application of article XIII in Nigeria. SC70 Doc. 27.3.5. Available at: https://cites.org/sites/default/files/eng/com/sc/70/E-SC70-27-03-05.pdf. Accessed 3 September 2020.

———. 2018b. Implementation report 2015–2017 Austria. Available at: https://cites.org/sites/default/files/reports/15-17Austria.pdf. Accessed 23 July 2020.

———. 2016a. Decisions of the conference of the parties to CITES in effect after its 17th meeting.

———. 2016b. The CITES species. Available at: https://cites.org/eng/disc/species.php. Accessed 26 July 2016.

———. 2016c. National laws for implementation of the convention. Available at: https://cites.org/sites/default/files/eng/com/sc/66/Inf/E-SC66-Inf-19.pdf. Accessed 26 July 2016.

———. 2016d. New implementation report format (formerly known as biennial report). Available at: https://cites.org/eng/node/43120. Accessed 11 November 2020.

———. 2016e. Resolution Conf. 17.6— Prohibiting, preventing, detecting and countering corruption, which facilitates activities conducted in violation of the convention. Available at: www.cites.org/sites/default/files/document/E-Res-17-06_0.pdf. Accessed 17 November 2020

———. 2013. CITES electronic permits toolkit version 2.0. Available at: https://cites.org/sites/default/files/eng/prog/e/cites_e-toolkit_v2.pdf. Accessed 11 August 2020.

———. 2013 [2016]. Resolution Conf. 16.7 (Rev. CoP17). Non-detriment findings. Available at: www.cites.org/sites/default/files/document/E-Res-16-07-R17_0.pdf. Accessed 3 September 2020.

———. 2010. Guidelines for a procedure to register and monitor operations that breed Appendix-I animal species for commercial purposes. Resolution Conf. 12.10 (Rev. CoP15). Available at: www.cites.org/sites/default/files/document/E-Res-12-10-R15.pdf. Accessed 10 August 2020.

———. 2007 [2010]. Resolution Conf.14.7 (Rev. CoP15) Management of nationally established export quotas. Available at: www.cites.org/sites/default/files/document/E-Res-14-07-R15_0.pdf. Accessed 3 September 2020.

———. 2005a. Albania biennial report—notification No. 2005/035. Available at: https://cites.org/sites/default/files/reports/03-04Albania.pdf. Accessed 23 July 2020.

———. 2005b. Notification No. 2005/035. Available at: https://cites.org/eng/node/1759. Accessed 11 November 2020.

———. 2002 [2019]. Resolution Conf. 12.8(Rev. CoP18). Review of significant trade in specimens of appendix-II species. Available at: https://cites.org/sites/default/files/document/E-Res-12-08-R18.pdf. Accessed 3 September 2020.

———. 2002. SC45 Doc. 21. Conservation of and trade in specific species.

———. 2000 [2019]. National reports. Available at: www.cites.org/sites/default/files/document/E-Res-11-17-R18.pdf. Accessed 26 August 2020.

———. 1995 [2013]. Resolution Conf. 10.21 (Rev. CoP16). Transport of live specimens. Available at: https://cites.org/sites/default/files/document/E-Res-10-21-R16.pdf. Accessed 11 December 2020.

———. 1992. Resolution Conf. 11.1 (Rev. CoP16), establishment of committees. Available at: https://cites.org/eng/res/11/11-01R16.php. Accessed 26 July 2016.

———. 1979. Bonn amendment to the text of the convention. Available at: https://cites.org/eng/disc/bonn.php. Accessed 10 December 2020.

———. 1973. Convention on international trade in endangered species of wild fauna and flora. Available at: www.cites.org/eng/disc/text.php#IX. Accessed 16 July 2020.

Cooney, R. and Abensperg-Traun, M. 2013. Raising local community voices: CITES, livelihoods and sustainable use. *Review of European, Comparative and International Environmental Law*. Special Issue—Focus on: CITES+40. 22 (3): 301–310.

Costa Rica. 2013. Biodiversity law: The legislative assembly of the republic of Costa Rica. Available at: www.faolex.fao.org/document/texts/biodiversitylaw. Accessed 15 December 2020.

Couzens, E. 2013. CITES at forty: Never too late to make lifestyle changes. *Review of European Community & International Environmental Law*. 22 (3): 311–323.

D'Cruz, N. and Macdonald, D. 2015. Clouded in mystery: The global trade in clouded leopards. *Biodiversity Conservation*. 24: 3505–3526.

De Roe, R., Melnychuk, N., Murray, D. and Plummer, R. 2016. Advancing the state of policy Delphi practice: A systematic review evaluating methodological evolution, innovation, and opportunities. *Technological Forecasts and Social Change*. 104, March: 78–88.

Di Minin, E., Laitila, J., Montesino-Pouzols, F., Leader-Williams, N., Slotow, R., Goodman, P. S. and Moilanen, A. 2015. Identification of policies for a sustainable legal trade in rhinoceros horn based on population projection and socioeconomic models. *Conservation Biology*. 29 (2): 545–555.

DLA Piper. 2015. Empty threat 2015: *Does the Law Combat Illegal Wildlife Trade? A Review of Legislative and Judicial Approaches in 15 Jurisdictions*. Washington, DC: DLA Piper and WWF.

———. 2014. *Empty threat: Does the Law Combat Illegal Wildlife Trade? An Eleven-Country Review of Legislative and Judicial Approaches*. Washington, DC: DLA Piper and WWF.

Dominica. 1990. Forestry and wildlife act. Chapter 60:02. Available at: www.dominica.gov.dm/laws/chapters/chap60-02.pdf. Accessed 7 August 2020.

European Commission. 2018. Joint staff working document: The EU special incentive arrangement for sustainable development and good governance ('GSP+') assessment of Armenia covering the period 2016–2017. Available at: http://trade.ec.europa.eu/doclib/docs/2018/january/tradoc_156537.pdf. Accessed 11 August 2020.

———. 2006. Commission Regulation (EC) 865/2006 laying down detailed rules concerning the implementation of Council Regulation (EC) No 338/97 on the protection of species of wild fauna and flora by regulating trade therein. Available at: www.legislation.gov.uk/eur/2006/865/adopted. Accessed 10 August 2020.

Felbab-Brown, V. 2018. *The Extinction Market: Wildlife Trafficking and How to Counter it. International Affairs*. London: Hurst & Company.

Fiji. 2017. Act No. 10 of 2017—an act to amend the endangered and protected species act 2002. Available at: http://extwprlegs1.fao.org/docs/pdf/fij171675.pdf. Accessed 11 August 2020.

Finnish Ministry of the Environment. 1995. *Nordic Research Project on the Effectiveness of International Environmental Agreements Draft Report*. Helsinki: Finnish Government.

Forest Research. 2020. Ramorum disease (*Phytophthora ramorum*). Available at: www.forestresearch.gov.uk/tools-and-resources/pest-and-disease-resources/ramorum-disease-phytophthora-ramorum/. Accessed 10 December 2020.

Foundation Franz Weber. 2019. Press release: Historic CITES decisions prevent the export of more than 30 baby elephants in Zimbabwe. Available at: https://africanelephantjournal.com/press-release-historic-cites-decisions-prevent-the-export-of-more-than-30-baby-elephants-in-zimbabwe/. Accessed 20 July 2020.

Fukushima, C., Mammola, S. and Cardoso, P. 2020. Global wildlife trade permeates the tree of life. *Biological Conservation*. 247: 180503: https://doi.org/10.1016/j.biocon.2020.108503.

Global Initiative to End Wildlife Crime. 2020a. Form and content of a possible protocol on the illicit trafficking of wildlife. Available at: https://endwildlifecrime.org/untoc-wildlife-protocol/. Accessed 17 November 2020.

———. 2020b. International coalition urges countries to #EndWildlifeCrime UN world wildlife day 2020. Available at: https://endwildlifecrime.org/wp-content/uploads/2020/07/PR-Wildlife-Crime_030320.pdf. Accessed 17 November 2020.

Government Gazette of the Republic of Namibia. 2008. No. 4190. Available at: https://cites.org/sites/default/files/projects/NLP/NamibiaControlledWildlifeProductsandTradeAct9.2008.pdf. Accessed 28 July 2020.

Government of Iceland. 2015. Importation of animals acts 1990 No 54, 16 March. Available at: www.government.is/library/04-Legislation/Act%20no%2054_1990-Importation%20of%20Animals%20Act-mai-2015.pdf. Accessed 11 August 2020.

Green, A. and The Center for Public Integrity (CPI). 1999. *Animal Underworld: Inside America's Black Market for Rare and Exotic Species*. New York: Public Affairs.

Guyana. 2016. Wildlife conservation and management bill 2016. Available at: http://dev.ultimate-dimensions.net/nredev/wp-content/uploads/2016/06/WILDLIFE-CONSERVATION-AND-MANAGEMENT-BILL-2016.pdf. Accessed 10 August 2020.

Harvey, R., Brien, M. L. Cherkiss, M. S., Dorcas, M., Rochford, M., Snow R. W. and Mazzotti, F. J. 2008. Burmese pythons in South Florida: Scientific support for invasive species management. Available at: http://edis.ifas.ufl.edu/pdffiles/UW/UW28600.pdf. Accessed 28 May 2009.

Hruby, D. 2020. He once trafficked in rare birds. Now, he tells how it's done. *The New York Times*. Available at: www.nytimes.com/2020/11/27/world/europe/rare-birds-macaws-trafficking-brazil.html. Accessed 1 December 2020.

Human Society International/Global. 2019. Momentous decision by CITES to limit trade in wild caught African elephants from Zimbabwe and Botswana for zoos to "exceptional circumstances" with independent oversight. Available at: www.hsi.org/news-media/cites-cop18-compromise-wild-caught-elephant-trade/. Accessed 20 July 2020.

Hutchinson, A., Stephens Griffin, N. and Wyatt, T. Under review. Speciesism and the wildlife trade: Who gets listed, delisted and uplisted in CITES? *International Criminology*.

Hyatt, A., Williamson, M., Coupar, B., Middleton, D., Hengstberger, S. and Gould, A. 2002. First Identification of a Ranavirus from Green Pythons (Chondropython viridis). *Journal of Wildlife Diseases*. 38: 239–252.

INBA. 2009. National wildlife trade policy review. Madagascar. Available at: www.inba.info_car.pdf. Accessed 14 December 2020.

Intergovernmental Science-Policy Platform on Biodiversity and Ecosystem Services (IPBES). 2019. IPBES global assessment summary for policymakers. Available at: www.ipbes.net/sites/default/files/downloads/summary_for_policymakers_ipbes_global_assessment.pdf. Accessed 7 May 2019.

International Air Transport Association (IATA). 2020. 2021 live animal regulations. Available at: https://store.iata.org/IEC_ProductDetails?id=9105-47&_ga=2.235785729.1262246004.1607673992-118588979.1607673992. Accessed 11 December 2020.

International Fund for Animal Welfare (IFAW) and International Union for the Conservation of Nature (IUCN). 2019. *Guidelines for the Confiscation of Live Wildlife*. Gland, Switzerland: IUCN.

International Institute for Sustainable Development (IISD). 2019. Earth negotiations bulletin. August. Available at: http://enb.iisd.org/vol21/enb21101e.html?utm_medium=email&utm_campaign=ENB%20Update%20%2030%20August%202019%20SW&utm_content=ENB%20Update%20%2030%20August%202019%20SW+CID_a7fade43500fa9f1ca5d10a7a4d2b3fa&utm_source=cm&utm_term=Read. Accessed 21 October 2019.

International Union for the Conservation of Nature (IUCN). 2020. Homepage. Available at: www.iucnredlist.org. Accessed 3 July 2020.

———. 2016. Categories and criteria. Available at: http://s3.amazonaws.com/iucn-redlist-newcms/staging/public/attachments/3097/redlist_cats_crit_en.pdf. Accessed 26 July 2016.

Jiménez-Bustamante, D. and Díaz Rentería, L. P. 2018. Laundering of illegal wild fauna in Mexico: Case study of a pair of desert monitors *Varanus griseus griseus* (Daudin, 1803). In Arroyo-Quiroz, I. and Wyatt, T. (eds.). *Green Crime in Mexico: A Collection of Case Studies*. Palgrave Studies in Green Criminology. London: Palgrave Macmillan: 127–136.

Kakabadse, Y. 2011. The fight against wildlife crime—enforcement v. corruption. *Environmental Policy and Law*. 41 (3): 123–126.

Karesh, W., Cook, R., Bennett E. and Newcomb, J. 2005. Emerging infectious diseases. Available at: www.cdc.gov/ncidod/EID/vol11no07/05-0194.htm. Accessed 20 May 2011.

Kimbrough, L. 2020. Conservationists set the record straight on COVID-19's wildlife links. *Mongabay*. Available at: https://news.mongabay.com/2020/03/conservationists-set-the-record-straight-on-covid-19s-wildlife-links/. Accessed 16 March 2020.

Legal Atlas. 2018. Home. Available at: www.legal-atlas.com. Accessed 23 July 2020.

Legifrance. 2020. Arrêté du 30 juin 1998 fixant les modalités d'application de la convention sur le commerce international des espèces de faune et de flore sauvages menacées d'extinction et des règlements (CE) n° 338/97 du Conseil européen et (CE) n° 939/97 de la Commission européenne. Available at: www.legifrance.gouv.fr/affichTexte.do?cidTexte=LEGITEXT000005626370. Accessed 5 August 2020.

Maher, J. and Wyatt, T. 2017. International trade in animals and animal parts. In Beirne, P., Maher, J. and Pierpont, H. (eds.). *International Handbook on Animal Abuse Studies*. Basingstoke: Palgrave Macmillan: 223–247.

Mallapaty, S. 2020. COVID mink analysis shows mutations are not dangerous—yet. *Nature*. 587: 340–341.

Marceau, J. 2019. *Beyond Cages: Animal Law and Criminal Punishment*. Cambridge: Cambridge University Press.

Margulies, J., Bullough, L., Hinsley, A., Ingram, D., Cowell, C., Goettsch, B., Klitgard, B., Lavorgna, Sinovas, and Phelps, J. 2019. Illegal wildlife trade and the persistence of "plant blindness". *Plants, People, Planet*. https://doi.org/10.1002/ppp3.10053.

Mitchell, R. 1996. Compliance theory: An overview. In Cameron, J., Werksman, J. and Roderick, P. (eds). *Improving Compliance with International Environmental Law*. Oxon: Earthscan: 3–28.

Mora. C., Tittensor, D. P., Adl, S., Simpson, A. G. B. and Worm, B. 2011. How many species are there on earth and in the ocean? *PLoS Biology*. 9 (8): e1001127. https://doi.org/10.1371/journal.pbio.1001127.

Munoz, M. A. C. 2016. The role of CITES in ensuring sustainable and legal trade in wild fauna and flora. In Elliott, L. and Schaedla, W. (eds). *Handbook of Transnational Environmental Crime*. Cheltenham: Edward Edgar: 433–442.

Naim, M. 2005. *Illicit: How Smugglers, Traffickers and Copycats Are Hijacking the Global Economy*. London: William Heinemann.

Nepal Law Commission. 2018. *An Act to Regulate and Control International Trade in Endangered Wild Fauna and Flora*. Available at: www.lawcommission.gov.np/en/archives/category/documents/prevailing-law/statutes-acts/an-act-to-regulate-and-control-international-trade-in-endangered-wild-fauna-and-flora. Accessed 13 August 2020.

Nhongo, K. 2019. Southern African nations threaten to leave global wildlife pact. *Bloomberg*. Available at: www.bloomberg.com/news/articles/2019-08-27/rhino-snub-draws-threat-of-africa-exit-from-global-wildlife-pact. Accessed 20 July 2020.

Nijman, V. and Shepherd, C. 2015. *Adding Up the Numbers: An Investigation into Commercial Breeding of Tokay Geckos in Indonesia*. Petaling Jaya, Selangor, Malaysia: TRAFFIC.

Nurse, A. 2016. Beyond the property debate: animal welfare as a public good. *Contemporary Justice Review*. Volume 19, Issue, Animals, Justice and the Law Part 2: 174–187.

Nurse, A. and Wyatt, T. 2020. *Wildlife Criminology*. Bristol: Policy Press.

Oldfield, S. 2003. Introduction. In Oldfield, S. (ed.). *Trade in Wildlife: regulation for Conservation*. London: Earthscan: xvii–xxvii.

Ortiz-von Halle, B. 2018. *Bird's Eye View: Lessons from 50 Years of Bird Trade Regulation & Conservation in Amazon Countries*. Cambridge: TRAFFIC.

Pacific Island Legal Information Institute. No date. Environmental protection—title 24—chapter 10. Available at: www.paclii.org/pw/legis/consol_act/ept24335/. Accessed 7 August 2020.

Rasmussen, S. 2020. Denmark to Cull 17 million mink after coronavirus mutation found. *The Wall Street Journal*. Available at: www.wsj.com/articles/denmark-to-cull-17-million-mink-after-coronavirus-mutation-found-11604592156. Accessed 17 November 2020.

Raymakers, C. 2002. Biodiversity and wildlife trade: From fauna and flora in trade to sanctions. In Anton, M., Dragffy, N., Pendry, S. and Young, T. R. (eds.) *Proceedings of the International Expert Workshop on the Enforcement of Wildlife Trade Controls in the EU, 5–6 Nov 2001*. TRAFFIC/IUCN.

Reeve, R. 2006. Wildlife trade, sanctions and lessons from the CITES regime. *International Affairs*. 82: 881–897.

———. 2002. *Policing International Trade in Endangered Species: The CITES Treaty and Compliance*. London: Earthscan.

Republic Moldova. 1995. Regnului animal. Available at: http://extwprlegs1.fao.org/docs/pdf/mol74856.pdf. Accessed 5 August 2020.

Robinson, J. E. and Sinovas, P. 2018. Challenges of analyzing the global trade in CITES-listed wildlife. *Conservation Biology*. 32 (5), 1203–1206. https://doi.org/10.1111/cobi.13095.

Rodgers, C. 2013. *The Law of Nature Conservation*. Oxford: Oxford University Press.

Sand, P. 2013. Enforcing CITES: The rise and fall of trade sanctions. *Review of European Community & International Environmental Law*. 22 (3): 251–263.

Scanlon, J. 2013. CITES at its best: CoP16 as a 'watershed moment' for the world's wildlife. *Review of European Community & International Environmental Law*. 22 (3): 222–227.

Schalow, F. 2015. Heidegger, the law of being and animal protection laws. *Ethics & the Environment*. 20 (2): 61–82.

Scheffers, B., Oliveira, B., Lamb, I. and Edwards, D. 2019. Global wildlife trade across the tree of life. *Science*. 366 (6461): 71–76. https://doi.org/10.1126/science.aav5327.

Science Daily. 2008. Python snakes, an invasive species in Florida, could spread to one third of us. Available at: www.sciencedaily.com/ releases/2008/02/080223111456.htm. Accessed 28 May 2009.

Sierra, Leone. 1972. The wild life conservation act 1972. Available at: http://extwprlegs1.fao.org/docs/pdf/sie41659.pdf. Accessed 11 August 2020.

Simard, S. 2012. Mycorrhizal networks: Mechanisms, ecology and modeling. *Fungal Biology Reviews*. 26: 39–60.

Sollund, R. 2019. *The Crimes of Wildlife Trafficking: Issues of Justice, Legality and Morality*. London: Routledge.

Sollund, R. and Runhovde, S. 2020. Responses to wildlife crime in post-colonial times. Who fares best? *The British Journal of Criminology*. 60 (4): 1014–1033. https://doi.org/10.1093/bjc/azaa005.

South Korea. 2014. Wildlife protection and management act. Available at: http://extwprlegs1.fao.org/docs/pdf/kor163043.pdf. 11 December 2020.

Statute of the Bahamas. 2016. Chapter 250A wildlife conservation and trade. Available at: http://extwprlegs1.fao.org/docs/pdf/bha78747.pdf. Accessed 7 August 2020.

Sudan. 2003. Laws of the new Sudan: Wildlife conservation and national parks act, 2003. Available at: http://extwprlegs1.fao.org/docs/pdf/ssd152460.pdf. Accessed 11 August 2020.

Swift, L., Hunter, P., Lees, A. and Bell, D. 2007. Wildlife trade and the emergence of infectious diseases. *EcoHealth*. 4 (25): 25–30.

The Economist. 2013. Polar bear politics: The Inuit take on world's greens and animal rightists Available at: www.economist.com/the-americas/2013/02/23/polar-bear-politics. Accessed 19 February 2020.

The Gambia. 2003. Biodiversity and wildlife act 2003. Available at: http://extwprlegs1.fao.org/docs/pdf/gam158129.pdf. Accessed 14 December 2020.

United Nations Food and Agricultural Organization (FAO). 2020. Establishing and linking global fish-trade information networks. Available at: www.fao.org/in-action/establishing-and-linking-global-fish-trade-information-networks/en. 16 August 2020.

———. 2012. Wildlife (conservation and security) act, 2012. Available at: http://extwprlegs1.fao.org/docs/pdf/bgd165019.pdf. Accessed 31 July 2020.

United Nations Office on Drugs and Crime (UNODC). 2020. *World Wildlife Crime Report: Trafficking in Protected Species*. Vienna: UNODC.

———. 2016. *World Wildlife Crime Report: Trafficking in Protected Species*. Vienna: UNODC.

———. 2000. United Nations convention against transnational organized crime definition. Available at: www.unodc.org/unodc/en/organized-crime/intro/UNTOC.html. Accessed 2 November 2020.

Vasquez, J. C. 2003. Compliance and enforcement mechanisms of CITES. In Oldfield, S. (ed.). *Trade in Wildlife: Regulation for Conservation*. London: Earthscan: 63–69.

Vincent, A. C. J., Sadovy de Mitcheson, Y., Fowler, S. L. and Lieberman, S. 2013. The role of CITES in the conservation of marine fishes subject to international trade. *Fish*. 14: 1–28.

Wandesforde-Smith, G. 2016. Looking for law in all the wrong places? Dying elephants, evolving treaties & empty threats. *Journal of International Wildlife Law & Policy*. 19 (4): 365–381.

Waterland, S., Vaughan, J., Lyman, E. and Jurisic, I. 2015. Illegal wildlife trade review, Malawi. Available at: www.lilongwewildlife.org/wp-content/uploads/IWT-Review-Malawi.pdf. Accessed 30 July 2020.

Webb, G. J. W. 2013. *Wildlife Conservation: In the Belly of the Beast*. Darwin, NT, Australia: Charles Darwin University Press.

Wijnstekers, W. 2018. *The Evolution of CITES: A Reference to the Convention on International Trade in Endangered Species of Wild Fauna and Flora, 11th Edition*. International Council for Game and Wildlife Conservation. Available at: www.cic-wildlife.org/wp-content/uploads/2018/05/the-evolution-of-cites-2018-11th-editionpart-1.pdf. Accessed 7 July 2020.

———. 2003. *The Evolution of CITES: A Reference to the Convention on International Trade in Endangered Species of Wild Fauna and Flora, 7th Edition*. Geneva: International Council for Game and Wildlife Conservation.

Wilkinson, K., Grant, W., Green, L., Hunter, S., Jeger, M., Lowe, P., Medley, G., Mills, P., Phillipson, J., Poppy, G. and Waage, J. 2011. Infectious disease of animals and plants: An interdisciplinary approach. *Philosophical Transactions of the Royal Society of London. Series B, Biological Sciences*. 366 (1573): 1933–1942.

Williams, V. and 't Sas-Rolfes, M. 2019. Born captive: A survey of the lion breeding, keeping and hunting industries in South Africa. *PLoS ONE*. 14 (5): 1–31. https://doi.org/10.1371/journal. pone.0217409.

Wilson, E. O. 2016. *Half Earth: Our Planet's Fight for Life*. New York: Liveright Publishing.

Wohlleben, P. 2016. *The Hidden Life of Trees: What They Feel, How They Communicate-Discoveries from a Secret World*. Vancouver: Greystone Books.

World Animal Protection. 2020. End the global wildlife trade. *Forever*. Available at: www.worldanimalprotection.org/take-action/end-global-wildlife-trade-forever. Accessed 10 December 2020.

World Health Organization (WHO). 2020. Coronavirus disease (COVID-19) pandemic. Available at: www.who.int/emergencies/diseases/novel-coronavirus-2019. Accessed 10 December 2020.

———. 2006. Neglected zoonotic diseases. Available at www.who.int/neglected_diseases/diseases/zoonoses/en/. Accessed 01 January 2020.

World Organisation for Animal Health (OIE). 2020. COVID portal—events in animals. Available at: www.oie.int/en/scientific-expertise/specific-information-and-recommendations/questions-and-answers-on-2019novel-coronavirus/events-in-animals/. Accessed 10 December 2020.

World Wildlife Fund (WWF). 2020a. Rhino. Available at: www.worldwildlife.org/species/rhino. Accessed 22 June 2020.

———. 2020b. Tiger facts. Available at: www.worldwildlife.org/species/tiger. Accessed 10 December 2020.

———. 2017. Unsustainable and illegal wildlife trade. Available at: http://wwf.panda.org/about_our_earth/species/problems/illegal_trade/. Accessed 01 February 2017.

Worrall, S. 2016. Meet the world' most expensive pet fish. *National Geographic.* Available at: www.nationalgeographic.com/news/2016/07/dragon-fish-most-expensive-arowana-emily-voigt/?cmpid=org=ngp::mc=crm-email::src=ngp::cmp=editorial::add=SpecialEdition_Escape_20200805&rid=720BDCA3E3D4E41A15BE807B71F01F25. Accessed 7 August 2020.

Wyatt, T. 2020a. Canada and the convention on international trade in endangered species of wild fauna and flora (CITES): Lessons learned on implementation and compliance. *Liverpool Law Review.* https://doi.org/10.1007/s10991-020-09267-8.

———. 2020b. Case study 2 — Indonesia. Available at: http://drtwyatt.weebly.com/uploads/2/4/3/1/24317071/indonesia_summary.pdf. Accessed 7 October 2020.

———. 2013a. The security implications of the illegal wildlife trade. *Journal of Social Criminology: Special Issue on Green Criminology.* Autumn: 130–158.

———. 2013b. *Wildlife Trafficking: A Deconstruction of the Crime, the Victims and the Offenders.* London: Palgrave Macmillan.

Wyatt, T. Friedman, K. and Hutchinson, A. Under review a. Are fish wild? *Animal Law Review.*

Wyatt, T., Johnson, K., Hunter, L., George, R. and Gunter, R. 2017. Corruption and wildlife trafficking: Three case studies involving Asia. *Asian Journal of Criminology.* 13: 35–55. https://doi.org/10.1007/s11417-017-9255-8.

Wyatt, T., Maher, L., Allen, D., Bruschi, N., Clarke, N. and Rook, D. Under review b. The welfare of the wildlife: How the legal and illegal trade debates ignore the wildlife. *Crime, Law and Social Change.*

Wyatt, T., van Uhm, D. and Nurse, A. 2020. Differentiating criminal networks in the illegal wildlife trade: Organised, corporate and disorganised crime. *Trends in Organized Crime.* https://doi.org/10.1007/s12117-020-09385-9.

Yeater, M. 2002. Enforcement and the CITES national legislation project. In Anton, M., Dragffy, N., Pendry, S. and Young, T. R. (eds.) *Proceedings of the International Expert Workshop on the Enforcement of Wildlife Trade Controls in the EU, 5–6 Nov 2001.* TRAFFIC/IUCN.

APPENDIX A

Delphi iterative survey round 1

I am seeking your help with my UK Arts and Humanities Research Council (AHRC) Leadership Fellowship funded project. My research aims to address the lack of empirical investigation of CITES implementation and compliance. It may be that the regulation in countries with species decline and lack of compliance can benefit from improved regulation in the form of legislation, implementation, and compliance mechanisms rather than or in addition to improved enforcement. In order to contribute new data to this discussion, I have analysed the academic and grey literature for each of the CITES member countries' publicly available wildlife trade legislation, and compliance data from CITES to understand what is known about the nature and extent of implementation of and compliance with CITES legislation.

Based upon this analysis, I am now conducting a "Delphi iterative survey" followed by semi-structured interviews to generate three case studies of lessons learned. The Delphi iterative survey will focus on what experts believe to be best practice and which countries have implemented and/or complied with CITES in ways that should be shared more widely. Responses are collated and then distributed again to all participants, so that they can see and comment on others' contributions or adjust their responses based upon other participants' answers.

The data are intended to provide evidence to member countries as to how their own practice can be improved and to inform other multilateral environmental agreements of approaches to implementation and compliance. To potentially inform improvements to CITES legislation, implementation, and compliance, I will share the three case studies with CITES members. Additionally, CITES staff and staff of other global environmental conventions and international NGOs will be invited to a final dissemination conference around World Wildlife Day in March 2020. The aim is to exchange knowledge between the wider environmental governance community on how we can better protect other species and the planet.

Please contact me with any questions and/or if you would like to attend the final conference: tanya.wyatt@northumbria.ac.uk or +44-191-227-4015.

- I agree to participate in the research, which has been explained to me in full. ☐
- I understand that the data collected will be used for publications and destroyed upon completion of the project. ☐
- I understand that my participation will be kept anonymous and confidential. ☐
- I understand that I can withdraw from the project at any time (up to the point at which research is submitted for publication), without giving a reason. ☐

The four components of implementation

Implementation of CITES has been the purview of the National Legislation Project (NLP), which examines whether member countries have measures in place to:

1. designate at least one Management Authority and one Scientific Authority;
2. prohibit trade in specimens in violation of the Convention;
3. penalize such trade; or
4. confiscate specimens illegally traded or possessed[1]

Component 1

Breakdown of CITES Authorities	
Separate EA, MA, SA	41
Same EA, MA, SA	17
No EA	85
Separate MA and SA	58
Same MA and SA	29
Same EA and MA	14
No SA	5
Overlap of EA, MA, SA	9
Same—Other	4

(EA = Enforcement Authority; MA = Management Authority; SA = Scientific Authority)

Although not required by CITES, many countries (98) have created an Enforcement Authority, which is tasked with preventing and detecting violations of national legislation by members of the public and businesses (rather than the MA's role to ensure there are mechanisms in place to meet the convention requirements). Only five member countries appear to not have a Scientific Authority, which means they have not fully implemented CITES. All seem to have a Management Authority. Of note is the overlap of authorities or that agencies serve as both the Management and the Scientific Authorities.

QUESTION 1

Should member countries be required to have an Enforcement Authority?
Strongly Agree   Agree   Neither   Disagree   Strongly Disagree

1 https://cites.org/eng/legislation/National_Legislation_Project

QUESTION 2

Should any Enforcement Authority be separate from the other authorities?
Strongly Agree Agree Neither Disagree Strongly Disagree

QUESTION 3

Are there examples of an authority structure that other member countries could learn from?

QUESTION 4

Do you have any other comments on CITES authorities?

Component 2

Countries which Prohibit violations of the convention	
Yes	103
No	80

My analysis looked for the language in the legislation that explicitly prohibits trade that violates CITES[2]. I found that apparently only 56% of member countries explicitly prohibit trade of wildlife in violation of the convention as required by CITES (this supports the Categories of the NLP).

QUESTION 5

What can be done to improve implementation of *prohibition* measures?

Ways to improve whether or not countries prohibit trade in violation of CITES					
Improvement or alternative	*Strongly Agree*	*Agree*	*Neither*	*Disagree*	*Strongly Disagree*
Introduce trade suspensions against those countries which have not prohibited trade in violation of the convention					

(*Continued*)

2 CITES legislation or review of the legislation is available in English for 112 member countries. I have working knowledge of Spanish and Russian, which comprised a further 20 countries. Therefore, Google translate was relied upon in 47 instances. In four instances, the text could not be translated or was unclear (three in Arabic; one in Somali was unclear at times).

Ways to improve whether or not countries prohibit trade in violation of CITES					
Improvement or alternative	*Strongly Agree*	*Agree*	*Neither*	*Disagree*	*Strongly Disagree*
Introduce reduced quotas against those countries which have not prohibited trade in violation of the convention					
Create a task force or working group to assist countries with this component					
More visibility on CITES website and in the media about which countries do and do not prohibit trade in violation of the convention					
Establish a system of country mentorships to have countries assist each other					

QUESTION 6

Are there other ways to improve whether or not countries prohibit trade in violation of the Convention?

QUESTION 7

Are there examples around prohibition in your countries or others that you know, which other countries could learn from?

Component 3

Penalties	
Fines	16
Fine/Prison	99
Prison	3
Unclear	13
Blank	52

In 28 per cent of the cases, I was unable to identify the penalty for CITES violations in member countries' legislation. This may be because I did not find the legislation (4); it was in a format that could not be translated through Google translate (4), or there was no explicit inclusion of penalties (42), though these may be found in other statutes. In a majority of cases, countries fine, imprison, or fine and imprison offenders.

QUESTION 8

What can be done to improve legislation on the *penalization* of violations?

Ways to improve whether or not countries penalize trade in violation of CITES

Improvement or alternative	*Strongly Agree*	*Agree*	*Neither*	*Disagree*	*Strongly Disagree*
Introduce trade suspensions against those countries which have not penalized trade in violation of the convention					
Introduce reduced quotas against those countries which have not penalized trade in violation of the convention					
Create a task force or working group to assist countries with this component					
More visibility on CITES website and in the media about which countries do and do not penalize trade in violation of the convention					
Establish a system of country mentorships to have countries assist each other					

QUESTION 9

Are there other ways to improve whether or not a country penalizes trade in violation of the Convention?

QUESTION 10

Are there examples around penalization in your countries or others that you know, which other countries could learn from?

Of note is that from my analysis I suggest that 41 countries appear to have penalties that meet the benchmark established by the United Nations Office on Drugs and Crime (4 years) to make violations a serious crime. Furthermore, it seems 27 countries have specific provisions for when corporations violate CITES.

QUESTION 11

Should 4-year penalties be made the standard for CITES legislation?
Strongly Agree Agree Neither Disagree Strongly Disagree

Component 4

Confiscation of illegally traded or possessed wildlife and the equipment to commit the crime	
Wildlife only	71
Both (wildlife/equipment)	47
Equipment	8
Evidence	1
None	1
Blank	55

Although confiscation of illegal wildlife and repatriating where possible is part of CITES, over 35 per cent of member countries appear to have no mechanisms or provisions in place in relation to confiscations[3].

QUESTION 12

What can be done to better support countries in regard to confiscations?

Ways to improve whether or not countries confiscate illegally traded and possessed wildlife

Improvement or alternative	*Strongly Agree*	*Agree*	*Neither*	*Disagree*	*Strongly Disagree*
Introduce trade suspensions against those countries which do not confiscate illegally traded or possessed wildlife					
Introduce reduced quotas against those countries which do not confiscate illegally traded or possessed wildlife					
Create a task force or working group to assist countries with this component					
More visibility on CITES website and in the media about which countries do and do not confiscate illegally traded or possessed wildlife					
Establish a global or regional network of zoos, aquariums, and gardens that assist					

3 This analysis does not include whether the wildlife are actually confiscated, but whether there is legislation addressing the issue. Member countries may choose to kill the wildlife, which is important, but beyond the scope of this study.

QUESTION 13

Are there other ways to support better implementation of confiscation measures?

QUESTION 14

Are there examples of best practice in terms of confiscations of illegal wildlife?

QUESTION 15

In particular, are there good examples of handling live confiscations?

OVERALL IMPLEMENTATION

Legislation

Type of Legislation			
Animal	3	Environmental Crime	2
Animal and Plant separate	1	Environmental Security	1
Blank	11	Forestry and Hunting	1
CITES-specific	48	Forestry and Wildlife	1
Conservation	6	Hunting	3
Conservation and Parks	1	Trade	2
Environmental	80	Wildlife	21
Environmental and Agriculture	1	Wildlife Trade	1

Most countries weave CITES legislation into environmental legislation, others into wildlife legislation. Forty-eight countries appear to have CITES-specific legislation.

QUESTION 16

Are there examples of best practice of a country that has woven CITES into existing or other legislation?

QUESTION 17

And are there other examples of stand-alone CITES legislation that can provide guidance on best practice?

Stricter Measures than CITES					
Conditions	Category	Yes	No	No Info	No Answer
	Trade	95	6		9
	Taking	86	6		18

(*Continued*)

Stricter Measures than CITES					
	Possession	88	8	1	13
	Transport	74	15	1	20
Prohibition	Trade	37	27		46
	Taking	49	23		48
	Possession	33	24		53
	Transport	23	26		61

The analysis of the member countries' legislation revealed that 12 countries appear to have legislation that is stricter than required; in particular, there is a stricter permit system in place. However, according to the biennial reports from member countries a large majority of member countries claim that they have stricter measures in place for conditions as well as many claiming they have stricter measures in place in terms of prohibition. This suggests that many countries perceive they are doing more than CITES requires, even though this may not be the case.

QUESTION 18

What can account for the discrepancy between the analysis of the legislation, the countries' perceptions of their efforts, and the continued decline of many species in trade?

Countries claim they have stricter measures in place					
Reason for discrepancy	*Strongly Agree*	*Agree*	*Neither*	*Disagree*	*Strongly Disagree*
Countries have misinterpreted CITES					
Countries are misrepresenting their implementation of CITES					
The legislation is stricter, but enforcement is lax or weak					
Trade as governed by CITES does not conserve species					
Overexploitation overwhelms the stricter measures					
Illegal trade overwhelms the stricter measures					

QUESTION 19

What stricter measures are the most effective in tackling illegal trade?
Permit conditions
Registering traders

Registering producers
All trade requiring a permit (Appendixes I and II; import and export)
Restricting conditions for taking
Restricting conditions for trading
Restricting conditions for possessing
Restricting conditions for transporting
Bans
Other

QUESTION 20

Does the listing of a country in Category 1 (fully implemented), 2 (partly implemented), and 3 (have not implemented) by the NLP accurately capture whether a country has implemented CITES?

Strongly Agree Agree Neither Disagree Strongly Disagree

QUESTION 21

What mechanism other than NLP categories might be a better indicator of implementation? Or what improvements could be made to the NLP?

Alternatives to the National Legislation Project

Improvement or alternative	*Strongly Agree*	*Agree*	*Neither*	*Disagree*	*Strongly Disagree*
More detailed breakdown of the four components, i.e., which component is lacking and how					
Metrics related to annual and biennial report submission linked to categories					
Add new components, i.e., standard level of penalty					
More visibility on CITES website and in the media about a country's category					
Make the NLP a project independent of CITES					
Metrics related to delistings, uplistings, and extinctions linked to categories					
Use victimization devices to measure implementation—these may include a combination of statistical records of violations (illegal trade) and testimonials (from human communities impacted by overexploitation and illegal trade and potentially from people representing the wildlife)					

From the post-2015 biennial reports and literature, member countries and scholars have identified the following hinder implementation; would you agree or disagree and are there other concerns not identified?

QUESTION 22

Difficulties In Implementing CITES					
Difficulty Identified	*Strongly Agree*	*Agree*	*Neither*	*Disagree*	*Strongly Disagree*
Tight deadlines for reporting					
Tight deadlines for implementing new resolutions					
Data asked for are difficult to collect					
Hard-to-implement resolutions such as on musical instruments, which generate added work and seizures					
Lack of data from some members makes implementation difficult					
Lack of finances					
Incomplete records with no scientific names					
Lack of human resources					
Awareness of relevant organizations is limited					
Lack of harmonization on registering facilities that captive-breed Appendix I species					
Disposal of confiscated and accumulated wildlife					
Not all resolutions and decisions can be transposed into community law as needed					
The regulation of illegal trade is compromised by concerns that it could adversely affect the legal trade, and thus the material interests of many member countries					
Appendix listings and resolutions are based on political/financial interests rather than scientific or other evidence					

QUESTION 23

What other difficulties are there to implementing CITES?

QUESTION 24

Constraints To Implementing CITES					
Constraint Identified	*Strongly Agree*	*Agree*	*Neither*	*Disagree*	*Strongly Disagree*
Coordinating multiple agencies involved in implementation					
Overly complicated import-export procedures					
International pressure					
Illegal trade					
Insufficient resources (material, financial, human, technical)					
Weak capacity (human and material) of authorities					
Legislative development					
No place for confiscated live wildlife					
Harmonization with overseas territories					
Corruption					

QUESTION 25

What other constraints are there to implementing CITES?

COMPLIANCE

Biennial Report Compliance 2003—present	
No of Reports	*No of Countries*
0	67
1	27
2	13
3	9
4	10
5	17
6	26
7	15

Thirty-six per cent of member countries have not completed any biennial reports. Sixty-three per cent of member countries have completed three or fewer. Twenty countries are identified by CITES as needing attention because of the failure to implement the convention in full, which is apparent because of their lack of compliance with the annual reporting requirement.

QUESTION 26

What can be done to encourage member countries to complete their biennial reports?

Ways to improve compliance with reporting requirements

Improvements	*Strongly Agree*	*Agree*	*Neither*	*Disagree*	*Strongly Disagree*
Introduce trade suspensions against those countries which do not report					
Introduce reduced quotas against those countries which do not report					
Create a task force or working group to assist countries with reporting					
More visibility on CITES website and in the media about which countries do and do not report					
Have a dedicated session at each Standing Committee meeting to complete reports					
Have a dedicated session at each Conference of the Parties meeting to complete reports					

QUESTION 27

Which component is most important in regard to implementation? Rank the following from most important to least important.

Prohibition of trade in violation of the Convention
Having a Management and a Scientific Authority
Penalizing violations
Having measures to confiscate illegally traded and possessed wildlife
Implementing the resolutions from the CoP
Implementing the decisions from the CoP
Other

QUESTION 28

Which component is most important in regard to compliance? Rank the following from most important to least important.

Annual reports being completed and submitted on time
Biennial reports being completed and submitted on time
Enforcing the legislation that is implemented
Enforcing the resolutions from the CoP
Enforcing the decisions from the CoP

QUESTION 29

Which country has a good approach to implementation of and compliance with CITES?

QUESTION 30

From which country could other members learn lessons?

Final thoughts and future directions

QUESTION 31

Do you have any other comments or thoughts?

QUESTION 32

Should there be a new international treaty or convention dealing specifically with wildlife crime and trafficking?

Strongly Agree Agree Neither Disagree Strongly Disagree

If so, what kind?

QUESTION 33

Or, should wildlife crime and trafficking be added to the Protocols of the United Nations Convention on Transnational Organized Crime (UNTOC)?

Strongly Agree Agree Neither Disagree Strongly Disagree

A bit about you

Gender
Female, Male, Other, Prefer not to Say
Organization
Government, Civil Society, Academia, Other
Region
Africa, Asia, Europe, Oceania/Pacific, North America, South America

APPENDIX B

Delphi iterative survey round 2

I am seeking your help with my UK Arts and Humanities Research Council (AHRC) Leadership Fellowship funded project. My research aims to address the lack of empirical investigation of CITES implementation and compliance. It may be that the regulation in countries with species decline and lack of compliance can benefit from improved regulation in the form of legislation, implementation, and compliance mechanisms rather than or in addition to improved enforcement. In order to contribute new data to this discussion, I have analysed the academic and grey literature for each of the CITES member countries' publicly available wildlife trade legislation and compliance data from CITES to understand what is known about the nature and extent of implementation of and compliance with CITES legislation.

On the basis of this analysis, I have conducted one round of a "Delphi iterative survey". The Delphi iterative survey focuses on what experts believe to be best practice and which countries have implemented and/or complied with CITES in ways that should be shared more widely. I have collated the 31 responses and now am sharing the experts' insights. This second round of the "Delphi iterative survey" is designed to expand and clarify the Round 1 responses by allowing participants to see and comment on others' contributions or adjust their responses based upon other participants' answers.

The data are intended to provide evidence to member countries as to how their own practice can be improved and to inform other multilateral environmental agreements of approaches to implementation and compliance. To potentially inform improvements to CITES legislation, implementation, and compliance, I will share the three case studies with CITES members. Additionally, CITES staff and staff of other global environmental conventions and international NGOs will be invited to a final dissemination conference around World Wildlife Day in March 2020. The aim is to exchange knowledge between the wider environmental governance community on how we can better protect other species and the planet.

Please contact me with any questions and/or if you would like to attend the final conference: tanya.wyatt@northumbria.ac.uk or +44-191-227-4015.

- I agree to participate in the research, which has been explained to me in full. ☐
- I understand that the data collected will be used for publications and destroyed upon completion of the project. ☐
- I understand that my participation will be kept anonymous and confidential. ☐
- I understand that I can withdraw from the project at any time (up to the point at which research is submitted for publication), without giving a reason. ☐

The four components of implementation

Implementation of CITES has been the purview of the National Legislation Project (NLP), which examines whether member countries have measures in place to:

1 designate at least one Management Authority and one Scientific Authority;
2 prohibit trade in specimens in violation of the Convention;
3 penalize such trade; or
4 confiscate specimens illegally traded or possessed.[4]

Component 1—Authorities

Although not required by CITES, many countries (98) have created an Enforcement Authority, which is tasked with preventing and detecting violations of national legislation by members of the public and businesses (rather than the MA's role to ensure there are mechanisms in place to meet the convention requirements). As such, Round 1 asked whether an Enforcement Authority should be required (27 of 31 participants strongly agreed or agreed) and should it be separate from the Scientific and Management Authorities (20 of 31 participants strongly agreed or agreed).

When asked "Are there examples of an authority structure that other member countries could learn from?" participants suggested the Criminal Assets Bureau Ireland, South Africa, UK Border Force, and the US Fish and Wildlife Service.

Would you agree with participants' comments that:

QUESTION 1

Independence is more important than separation in regard to authorities.
Strongly Agree   Agree   Neither   Disagree   Strongly Disagree

4 https://cites.org/eng/legislation/National_Legislation_Project

QUESTION 2

Defining roles and responsibilities of the all (three) authorities is critical.
Strongly Agree Agree Neither Disagree Strongly Disagree

QUESTION 3

Authorities should be separate, but they should be co-located.
Strongly Agree Agree Neither Disagree Strongly Disagree

Component 2—Prohibition of Trade

My analysis looked for the language in the legislation that explicitly prohibits trade that violates CITES[5]. I found that apparently only 56% of member countries (103) explicitly prohibit trade of wildlife in violation of the convention as required by CITES (this supports the Categories of the NLP).

Participants answered that of the options given, "Create a task force or working group to assist countries with this component" and "Establish a system of country mentorships to have countries assist each other" were the best ways to improve implementation (29 of 31 strongly agree or agree). This was followed by "Introduce trade suspensions against those countries which have not prohibited trade in violation of the convention" and "More visibility on CITES website and in the media about which countries do and do not prohibit trade in violation of the convention" (24 of 31 strongly agree or agree). The least supported option was "Introduce reduced quotas against those countries which have not prohibited trade in violation of the convention" (16 of 31 strongly agree or agree).

Other suggestions were not given, but participants pointed out that looking at prohibition of trade without also looking at penalties and judicial action gave a simplistic overview. Undoubtedly, these elements are linked, but in terms of legal implementation, I maintain these are distinct components that can be, at least in part, assessed independent of one another.

Participants highlighted that prohibition of trade may only happen for native species that are CITES-listed.

QUESTION 4

One overlooked problem with countries regulating trade in CITES species is that members only *prohibit* trade in native species.

5 CITES legislation or review of the legislation is available in English for 112 member countries. I have working knowledge of Spanish and Russian, which comprised a further 20 countries. Therefore, Google translate was relied upon in 47 instances. In four instances, the text could not be translated or was unclear (three in Arabic; one in Somali was unclear at times).

Strongly Agree Agree Neither Disagree Strongly Disagree

QUESTION 5

One overlooked problem with countries regulating trade in CITES species is that members only *enforce* CITES restrictions for native species.
Strongly Agree Agree Neither Disagree Strongly Disagree

QUESTION 6

Stricter domestic measures are good mechanisms for preventing trade in wild-taken specimens of nationally endangered species.
Strongly Agree
Agree
Neither
Disagree
Strongly Disagree

Component 3—Penalization of trade violations

In 28 per cent of the cases, I was unable to identify the penalty for CITES violations in member countries' legislation. This may be because I did not find the legislation (4); it was in a format that could not be translated through Google translate (4), or there was no explicit inclusion of penalties (42), though these may be found in other statutes. In a majority of cases, countries fine (16), imprison (3), or fine and imprison offenders (99).

Participants answered that of the options given, "Create a task force or working group to assist countries with this component" (27 of 31 strongly agree or agree) was the best way to improve legislation on the penalization of violations. This was followed by "Establish a system of country mentorships to have countries assist each other" (26 of 31 strongly agree or agree). This was followed by "More visibility on CITES website and in the media about which countries do and do not penalize trade in violation of the convention" (22 of 31 participants strongly agree or agree), then "Introduce trade suspensions against those countries which have not penalized trade in violation of the convention" (19 of 31 participants strongly agree or agree), and the least supported option was "Introduce reduced quotas against those countries which have not penalized trade in violation of the convention" (18 of 31 strongly agree or agree).

Round 1 participants made a number of interesting suggestions as to other ways that violations of CITES could be penalized.

QUESTION 6

Asset forfeiture and/or confiscation is an underused penalty for CITES violations.

Strongly Agree Agree Neither Disagree Strongly Disagree

QUESTION 7

Importing countries could respond more strongly by not allowing countries with poor implementation of or compliance with CITES to import CITES species.

Strongly Agree Agree Neither Disagree Strongly Disagree

Component 4—Confiscation of illegally traded or possessed wildlife

Although confiscation of illegal wildlife (live and products) and repatriating where possible is part of CITES, over 35 per cent of member countries appear to have no mechanisms or provisions in place in relation to confiscations[6].

Participants felt that of the options given the following could support countries in regard to confiscations. First "Create a task force or working group to assist countries with this component" (29 of 31 participants strongly agree or agree). Then, "Establish a global or regional network of zoos, aquariums, and gardens that assist" (22 of 31 participants strongly agree or agree). Followed by "Introduce trade suspensions against those countries which do not confiscate illegally traded or possessed wildlife" assist' (21 of 31 participants strongly agree or agree), then "More visibility on CITES website and in the media about which countries do and do not confiscate illegally traded or possessed wildlife" assist' (20 of 31 participants strongly agree or agree). The least supported option was "Introduce reduced quotas against those countries which do not confiscate illegally traded or possessed wildlife" assist' (14 of 31 participants strongly agree or agree).

Participants shared additional options to improve implementation of the confiscation component of CITES as well as additional challenges.

QUESTION 8

Countries should consider implementing mechanisms for cost recovery in cases where any type of confiscations occur.

6 This analysis does not include whether the wildlife is actually confiscated, but whether there is legislation addressing the issue. Member countries may choose to kill the wildlife, which is important, but beyond the scope of this study.

Strongly Agree Agree Neither Disagree Strongly Disagree

QUESTION 9

Countries should consider implementing mechanisms for asset forfeiture in cases where any type of confiscations occurs.
Strongly Agree Agree Neither Disagree Strongly Disagree

QUESTION 10

Confiscated wildlife (live or products) re-entering the trade is an important problem to address.
Strongly Agree
Agree
Neither
Disagree
Strongly Disagree

QUESTION 11

In particular, are there good examples of handling confiscated wildlife products (i.e. evidence procedures, storage, etc.)?

Overall implementation

One participant said that overall CITES does conserve species, but enforcement (weak or lax) stymies success.

QUESTION 12

Do you agree with that statement?
Strongly Agree Agree Neither Disagree Strongly Disagree

QUESTION 13

Registering keepers of certain specimens is another stricter measure that would aid in tackling illegal wildlife trade.
Strongly Agree Agree Neither Disagree Strongly Disagree
Keeping close watch on certain companies is another stricter measure that would aid in tackling illegal wildlife trade.
Strongly Agree Agree Neither Disagree Strongly Disagree
Only 11 of 31 participants strongly agreed or agreed that the categories of the NLP accurately capture whether a country has implemented CITES. Ten participants chose neither agree nor disagree.

Twenty-seven of 31 participants strongly agreed or agreed that a "More detailed breakdown of the four components, i.e., which component is lacking and how" would improve the NLP.

Other suggestions were to:

QUESTION 14

Add a metric to the NLP about the number of successful prosecutions.

Strongly Agree Agree Neither Disagree Strongly Disagree

QUESTION 15

Add a metric to the NLP about the penalties allowed.

Strongly Agree

Agree

Neither

Disagree

Strongly Disagree

The main difficulties and the main constraints for implementation of CITES that were identified (as well as verifying the biennial reports) were insufficient financial and human resources and ongoing corruption and lack of political will. Furthermore:

QUESTION 16

Inconsistency between parties to the treaty is perhaps the biggest obstacle to its successful implementation.

Strongly Agree Agree Neither Disagree Strongly Disagree

Compliance

Thirty-six per cent of member countries have not completed any biennial reports. Sixty-three per cent of member countries have completed three or fewer. Twenty countries are identified by CITES as needing attention because of the failure to implement the convention in full, which is apparent because of their lack of compliance with the annual reporting requirement.

In order to improve upon this, participants felt that "a task force or working group to assist countries with reporting" was the best of the ideas proposed. Some participants commented that positive encouraging mechanisms should be used rather than punishments. However, one participant recommended:

QUESTION 16

To improve compliance exclude the non-complaint member from activities until they become compliant.

Strongly Agree Agree Neither Disagree Strongly Disagree

QUESTION 17

CITES should produce (as has been done in the past) a full report on non-compliance to share with all members.

Strongly Agree Agree Neither Disagree Strongly Disagree

QUESTION 18

The issue is not one of the quantity of reporting, but on the *quality* of reporting, which needs to be improved.

Strongly Agree
Agree
Neither
Disagree
Strongly Disagree

QUESTION 19

Any other comments are gladly received

QUESTION 20

I participated in the first round of this Delphi Iterative Survey
YES NO

A bit about you

Gender
Female, Male, Other, Prefer not to Say
Organization
Government, Civil Society, Academia, Other
Region
Africa, Asia, Europe, Oceania/Pacific, North America, South America

INDEX

Note: Page numbers in *italics* indicate a figure and page numbers in **bold** indicate a table on the corresponding page.